Springer Biographies

The books published in the Springer Biographies tell of the life and work of scholars, innovators, and pioneers in all fields of learning and throughout the ages. Prominent scientists and philosophers will feature, but so too will lesser known personalities whose significant contributions deserve greater recognition and whose remarkable life stories will stir and motivate readers. Authored by historians and other academic writers, the volumes describe and analyse the main achievements of their subjects in manner accessible to nonspecialists, interweaving these with salient aspects of the protagonists' personal lives. Autobiographies and memoirs also fall into the scope of the series.

Neal Brown

Northern Lights and Rocket Flights

My Years at Poker Flat Research Range

Neal Brown
c/o Kristopher Brown
New York, NY, USA

ISSN 2365-0613 ISSN 2365-0621 (electronic)
Springer Biographies
ISBN 978-3-032-14597-0 ISBN 978-3-032-14598-7 (eBook)
https://doi.org/10.1007/978-3-032-14598-7

This Springer imprint is published by the registered company Springer Nature Switzerland AG
The registered company address is: Gewerbestrasse 11, 6330 Cham, Switzerland

Editor's Note

When he died in 2021 at age 82, Neal B. Brown left behind his wife of 40 years, Fran Tannian; his children Kristopher Brown, Melody Brown Burkins, and Nathaniel Brown; his stepsons Steven Sweet and Michael Sweet; and six grandchildren. He also left 1100 pieces of writing that he had intended to craft into a history of the Poker Flat Research Range during his tenure there. These tidbits include anecdotes of triumphs and mishaps, pranks, and detailed histories of the evolution of literally rocket science as it was practiced at the only university-owned rocket range in the world, the only high-latitude rocket range in the United States, and the largest land-based rocket research range in the world. There are many stories of doing things "the Alaskan way": getting it done regardless of the obstacles as Poker Flat evolved over the years. Neal Brown was the supervisor and director of Poker Flat Research Range from 1971 to 1989, leading it from near–elimination through major expansions to what it is today: a premier facility not only for launching sounding rockets but conducting other specialized research. The range now incorporates rocket assembly and launching capabilities, telemetry receiving stations, and ground-based diagnostics needed for launch decisions concerning space, aeronomy, and atmospheric science experiments. Ground-based instrumentation allows monitoring of auroral activity, magnetic storms, ionospheric perturbations, and other space disturbances in real time.

Neal's oldest son, Kristopher, contacted me in 2022 to ask if I would attempt to craft Neal's snippets into a book. I had worked for years as the production editor for the University of Alaska Press, and then for 5 years as the public information officer for the University of Alaska Fairbanks Geophysical Institute. I knew Neal and Fran socially and professionally and had spent time at Poker Flat. I had worked as a free-lance editor and book designer for 30 years. So it seemed as if I should have been well qualified for the task.

The job was much more substantial than I anticipated. Kris's request was to focus only on writings about Poker Flat, to create a history of the range that would mainly begin after the events in T. Neil Davis's wonderful book *Rockets Over Alaska: The Genesis of Poker Flat*. After doing an inventory of the material with the help of my able assistant Ellen Bohman Mitchell, I set aside Neal's strictly personal writings, then organized all the writings about Poker Flat into roughly chronological order. When it became clear that structure did not work, I developed the hybrid organization that you see here.

This book is almost entirely Neal's words, edited (sometimes heavily) for clarity and consistency. The organization of the words (and any blame attached) is mine. Chapter 1 gives an account of Neal's early life and how he arrived at the University of Alaska Fairbanks and eventually became the first supervisor of Poker Flat. Chapter 2 consists of Neal's exploration of important milestones before the establishment of Poker Flat. Chapter 3 briefly summarizes the immediate cause of the first installation of a rocket launcher at Poker Flat and some associated other facilities, up until Neil Davis chose Neal Brown as the first full-time supervisor. Chapter 4 looks at Neal's first days, months, and years as manager of Poker Flat, lessons he learned, and some of the early improvements. Chapter 5 takes up the improvements theme to discuss some of the many additions to infrastructure, improvements in safety, and other changes to the facility. Chapters 6, 7, and 8 are mostly chronological stories about memorable rocket launches and other events through the end of the 1980s, when Neal retired from Poker Flat. Neal's immediate postretirement is briefly summarized in Chap. 9.

Sprinkled throughout the book are side stories. These are vignettes that don't always fit with the main narrative but that are fun, enlightening, or otherwise worthwhile. Also included are some photos from Neal's collection and others graciously contributed by his close friend and work collaborator Daniel Osborne. In this I am imitating Neil Davis in *Rockets Over Alaska*, who wrote, "In this autobiographical account I include anecdotes of peripheral or insignificant nature. I hope I can justify inclusion of these anecdotes on the basis of their entertainment value, and that some of this semirelevant material helps portray the times and the overall scientific, political, and philosophical setting in which this far north facility began." Neil Davis was a man of humor and insight, a towering intellect, and the driving force behind the initial establishment of Poker Flat.

This book represents one person's memories of events that happened in many cases 50 or more years ago. While I have attempted to verify facts, this account cannot be taken as absolute truth. Other people may well remember things differently.

I am greatly indebted to Neil Davis for the background and fact-checking his *Rockets Over Alaska* book provided, especially for his appendix list of Poker Flat Research Range launches from 1969 to 2005, to which Neal Brown contributed. More up-to-date launches are listed at https://www.pfrr.alaska.edu.

I am also grateful to Kris Brown for trusting me with his father's writings and giving me ample time and freedom to sculpt this narrative out of the raw material Neal left behind. Daniel Osborne generously reviewed the manuscript. Rachel Fudge did her usual professional job of copyediting. Any remaining errors of fact or differences of interpretation should be attributed to the vagaries of human memory and the shortcomings of the editor.

Sue Mitchell

Prologue: The Shining

Nights up North with the Aurora Borealis and the Rocket Men of Poker Flat

Grant Sims

I live in Anchorage. My friends 400 road miles to the north in Fairbanks call it Los Anchorage, and they say the nicest thing about Los Anchorage is that it's so close to Alaska. I used to live near Fairbanks myself, in a little town called North Pole, which I remember primarily because (1) my address convinced old acquaintances in the Lower 48 that I was no longer serious about life; (2) North Pole often recorded the coldest daily temperatures in North America; and (3) the area of interior Alaska in which both North Pole and Fairbanks are located is the best possible zone in the world from which to watch the aurora borealis, those brilliant ionospheric light shows we call the northern lights.

Visitors to interior Alaska don't see much aurora activity because they come mostly during summer, when it never really gets dark. During the dead of winter, of course, it never really gets light in the interior. Last year in Fairbanks you could see auroras on 243 nights. In Anchorage, just those few hundred miles south, aurora nights numbered fewer than 50. And most of those were obscured by the city's heavy winter cloud cover and its metropolitan city lights.

Fairbanks is the place for Alaska's aurora buffs and brains—especially the scientists from the University of Alaska's Geophysical Institute, which runs the Poker Flat Research Range. Poker Flat is 7000 acres of spruce bogs and radar and rocket gadgetry located 30 miles outside of Fairbanks, up a steep and icy little road in the Chatanika River Valley. The scientists there know all about such things as how auroras are really sparks from a 93-million-mile-long electromagnetic generator. And they say that when they get the generator all figured out, the auroral sparks will help teach us how to harness nuclear fusion and solve the world's energy problems forever.

In mid-January, with Fairbanks in 21 h of darkness daily, I drove north past Mount McKinley [Denali] toward the auroral zone. I planned to spend 5 days at Poker Flat, and my visit would begin with some spectacular though man-made lights. The scientists were going to launch a rocket right up into the aurora.

It was 40 below, and the Poker Flat people were using an oil heater to keep the rocket warm so it would ignite when the time came. Warmth radiated from the metal ducts in clouds of vapor that rolled down the sides of the rocket and spread out over the snow. The rocket stood there in whiteness like a 37-foot icicle. A red beacon blinked on a nearby meteorological instrument tower, and cold blue-white light pooled around the small command post beneath it. Today the bunker was cement, but 15 years ago, when this testing ground was built, the only protection against an accidental explosion was a wood shack surrounded by a wall of 55-gallon drums full of frozen water.

This particular launch had gone into its final countdown. So far it had been postponed eight times, mainly because the temperature was too cold for the rocket-recovery aircraft to fly. I had sat through most of these postponements, and my 5-day January visit had extended to 2 weeks—past Super Bowl Sunday and into the dead of the arctic winter. Now, with little more than an hour before launch, I was speeding away from the rocket, flying at 50 miles an hour over packed snow in a blue GMC pickup, aimed at a series of switchbacks up a mountain. It was 4:55 a.m. My driver was the research range supervisor, Neal Brown, a big man, 45 years old, loose, amiable, decisive, and heavy with the right foot.

"What does 'T' mean?" I asked him. He glanced, startled, and asked, "What time is it?"

"Two minutes till five," I replied.

He drove faster, fishtailing. "Time to launch," he said. "Pardon. 'T' means time to launch."

His truck headlights flashed a swath, picking out diamonds on the hard snow. Away from the brilliance of the command post, it was one of those crisp, blue-sheen nights, with rime ice luminescent on the stunted spruce. Overhead was a fair quivering of pearl aurora, the stuff that had kept Neal at this post for the past 12 years.

The aurora, Neal had told me, is a plasma. A large percentage of the known matter in the universe is plasma, yet the only place it is found naturally near the surface of the Earth is in the aurora and in the highest levels of the atmosphere. Plasma is important to humankind for a lot of reasons, but scientists get especially interested because an understanding of plasma is of fundamental importance to astronomy and astrophysics.

That's basic science, though, and these days basic science doesn't pay for rockets. Happily, research into the northern lights also has its practical side. Scientists working on plans to put nuclear-fusion-powered electric plants in space want to know how to protect them from the hot auroral plasmas of the magnetosphere. Other researchers want to find ways to stop auroral interference that can scramble radios and radar and send multimillion-dollar satellites careening out of control in space. They would also like to stop aurora-induced electrical currents from blowing transformers and accelerating electrolytic corrosion in the Trans-Alaska Pipeline.

But firsthand study of the lights is tricky. The zone, from 50 to 350 miles up, is too high for airplanes, too low for satellites. To get to it, scientists pack as many instruments as possible into a cylindrical metal payload and blast it off within a rocket. The rocket travels through the zone for a few minutes, gathering data that can take days, months, even years to analyze. At costs ranging from thousands to millions of dollars a whack, more than 200 such rockets have been launched from Poker Flat. The current launch, which we were waiting for, was being sponsored by NASA, and it was designed to determine just how fast electrons accelerate into the atmosphere during an auroral bombardment. Driving along, Neal said the experiment might provide one more formula, "one more bucket of cement for the foundation" on which to build the solution to the mysteries of astrophysics.

The blue GMC skittered around one last climbing curve and stopped at a white ridgetop dome capped by a large Plexiglas bubble. Neal called it an optics site, "because it's more than an observatory." He parked, left the engine running, motioned for me to follow, and poked his head inside the dome's front door.

"Clif?"

Clif wasn't there.

"He'll materialize." Neal looked apologetic. "I hate to drop you like this but …" An intercom speaker inside the dome crackled. "Gentlemen," said a firm female voice, "we are at T-minus 50 min and counting." Neal yanked the wolf fur ruff of his parka hood tight around his face and left.

I went inside. Shadows on the floors and walls of the dome jittered in time to the movements of red-and-blue lights from a floor-to-ceiling panel of television monitors, instrument faces, indicator bulbs, buttons, switches, needles, and dials.

I fumbled through a dark doorway off the main room, flipped a light switch, and found a kitchen. Setting my camera gear on a table, I took off my gloves, filled a teakettle with water, and waited for the whistle. In the other room, the intercom crackled: T-minus 45 min. I slipped on my gloves, cradled a Styrofoam cup of Brim, and walked outside to a rear deck that Neal had told me would provide the best vantage for photographing the liftoff.

From the deck I could see maybe 5 miles down and 10 miles up the Chatanika River Valley and a couple of miles across the valley to a ridge—the White Mountains north of Fairbanks. The sky was clear; the Big Dipper and the North Star were prominent. The aurora tonight was dim and scattered, and only occasionally and weakly did the lights move with their well-known devilish little darlings and dancings.

On a great night, the aurora looks like rampant rainbows. The lights slash and stab in blades and needles of green and violet; they shear off into isolated, red-feathered explosions, and then dance back into formation where they hang like a sedate but quivering final curtain.

Aristotle described the lights as brands, gulfs, abysses, and torches. Pliny, in reporting stories of armies clashing in the sky, theorized that the heavens had merely suffered a flash fire, which was "no extraordinary thing, and it has often been seen when the clouds are exposed to great heat."

Auroras were rare in what was then the civilized world, so Nero might be excused for hustling an army off to chase down an auroral glow, thinking his colony of Ostia was on fire—a mistake repeated by the Danish army in 1709. France's Henry III noted in his journal that in the autumn of 1583 his people made pilgrimages by the tens of thousands to holy shrines when the night skies were filled for several days with clashing armies, fire, bloody lances, and heads separated from trunks. In the far north, familiarity deprived the aurora of its terror.

One of the most unusual recent beliefs about the aurora emerged in turn-of-the-century Alaska, where a feverish rumor spread among hapless gold seekers that the northern glows were emitting from a massive circumpolar radium deposit. Bard Robert Service bit his tongue and wrote, "It's mine, all mine—and say! if you have a hundred plunks to spare/I'll let you have the chance of your life, I'll sell you a quarter share."

Standing on the deck of the optics site, I could see the red beacon of the instrument tower below, the white rocket piercing through its pool of bluish light, and the heat vapors spilling on the snow. Overhead, the night glow remained wispy, with faint spangles and weak electrical flickering spread thinly across the entire sky, a milky shimmer against the stars. In thc cold, dry silence I could hear snow squeak as someone walked outside the bunker a quarter of a mile away. I was getting cold. My beard and mustache were encased in ice, and in my plastic cup the remaining coffee was frozen hard as porcelain.

Coffee was one ingredient of an analogy Neal Brown used to explain to me why several "different" aurora times and intensities are really the same. "Pour cream into coffee," he said. "The cream makes a pattern; but the more you stir, the less the pattern there is, until there is no pattern at all. Still, you have the same amount of cream." So it is that while aurora energy can remain more or less constant from dusk to dawn, the lights seem to begin with early-evening "quiet arcs" that build to the midnight crescendo of "rayed curtains." They then fade into a scattering of individual rays and puffy "pulsating auroras" that look like galactic nebulae. It's all the same stuff, just the auroral cream mixing with ionospheric coffee.

Even at the height of an auroral display, we can see only a small portion of the light that's created. As I stood on the deck, the sky above me was actually alive with a violence of infrared and intense blue that I couldn't see.

Neal Brown has instruments to see all those invisible lights. He can sit at infrared scanners and elaborate optical spectrophotometers. The lights and the forces that create them are what intrigue Neal now, but it wasn't that way at first. As a 24-year-old newly graduated physics student, he went to work in 1961 for NASA to help solve rocket reentry heating problems. It so happens that rockets and meteorites burn up in the same atmospheric zone in which the auroras occur. Neal got sidetracked and wound up doing his master's thesis on auroral spectroscopy—the study of the spectrum of auroral colors and how their intensities and behaviors can be understood.

Ironically, he is color-blind.

At T-minus 10 min, activity on the intercom began picking up. Two dozen people were out there at various microphones, going over a checklist:

"FAA 10-min alert?"
"Check."
"Roadblocks deployed?"
"Check."
"TM bus voltage?"
"Twenty-eight volts."
"System current?"
"Three amps."

The voices would continue now, steadily, until blastoff. Unless, as had happened so far eight times, someone said, "Hold." Then there would be a silence of maybe 10 s while whoever ordered the hold pushed buttons or checked circuits or did whatever he had to do to make the right thing happen. Then the voice of the launch coordinator would come on: "How long is the hold?"

If it was a scrub, the team would break up, some heading for the bunkhouse, some bribing away their disappointment over a couple of consolation beers at the Chatanika Inn, 2 miles down the road. Some of them drive down to Fairbanks for a good meal, a quiet night of hotel oblivion, or just for a wild diversion at the Lonely Lady strip joint out at the west end of town.

During my 2 weeks of wandering in and out of the Poker Flat facilities, there were three groups present. One was me. The second consisted of transient scientists who come in for one-shot rocket experiments. They stay for a few days or a few weeks, however long it takes to get their rocket off, and then they go home to White Sands or Washington, DC, or Berkeley. Most of them make this midwinter trip to the frozen north an annual thing, and they have evolved some degree of savvy about when and where to eat and sleep, and how to draw straws or pitch pennies before driving to Fairbanks so that one man will stay sober for the cold drive back.

Among the third group—the dozen year-rounders—Neal Brown is dean. Neal reminds me of those biathlon competitors who can ski a frantic race and then, within seconds, breathe some special thoughts that cut their heart rates by half so that they can pull a calm, steady trigger. Each night of my visit, Neal would race the clock to blastoff, flying over the snow in the blue pickup, stopping, talking, deciding, moving on. In the midst of it all, he would come into his office fast, stomp the snow from his boots, flop into a swivel chair, sign twice, conduct another patient and articulate 45-min installment of our interview, sign twice more, and leave. He would never take off his coat, and he would always leave the pickup running. With this particular launch delayed and redelayed, he had been at it without wincing for 2 weeks, while I sleepwalked in his wake.

Not everyone is suited for the work here; you have to keep your wits. As you walk into the headquarters at Poker Flat, you are confronted by a rocket mounted just inside the door. Part of the rocket is cut away, and inside it is a beautiful woman who startles you with a wink as you pass. The same woman, who is a holograph, blows a kiss as you leave. Beyond the woman in the rocket is a full-size Darth Vader who sounds like James Earl Jones talking through a toilet-paper tube. Beside Darth Vader is R2-D2, who during my visit had not been reprogrammed since Christmas; it zipped around the headquarters bleeping "Jingle Bells."

In their isolation, the Poker Flat people obviously have their fun. It's always been this way. The facility's name is borrowed from a Bret Harte short story ("The Outcasts of Poker Flat") about a bunch of misfits cast into a frozen wilderness by a self-righteous citizenry. Then there are the research range's bunkhouse bathrooms, labeled "Poker" and "Pokee." There are randy jokes taped to walls, a framed montage of photographed horseplay, and a drink called Geofizz, over which to tell stories.

Like the one about the time a site review committee came upon a group of range personnel skinning a freshly killed moose as it hung from one of the rocket launchers.

Or the time the Poker Flat scientists contracted with the Los Alamos Scientific Laboratory (LASL) to launch a series of rockets into the upper atmosphere to release barium clouds. Bored with such mission names as ICECAP and CARIBOO, the scientists dubbed the new multiple-launch mission OOSIK—the Alaska Native name for a certain part of a male walrus's anatomy. A chastised but grinning Neal Brown who wrote in a subsequent mission report that "since the LASL administration was not too familiar with walrus anatomy or the Eskimo language, nor were they very tolerant of our attempt at humor, OOSIK now exists on the formal records as the acronym for Optical Observations of Substorm Induced Kinetics."

The boys at Poker Flat undertook three more barium-release missions for LASL, all coded with the innocuous names of South American birds.

At T-minus 3 min, I could feel my heartbeat in my temples and fingers. In 2 weeks of scrubs, the countdown had not made it this far. The intercom activity was frantic. I had been rehearsing with my Nikon. Its tripod was set up outside in the right spot, but I couldn't take the camera into the weather too soon or it would freeze.

Finally, at T-minus 2 min, I zipped my down coat, pulled on the hood, tucked the Nikon into a coat pocket, and walked outside.

For several moments I simply stood, staring at the rocket, wondering what it would sound like, how it would look. I had hoped to be here for one of the barium launches, in which a single pound of the chemical released at high altitude will create an artificial aurora 100 miles in diameter. Such a launch is called a tracer shot. The barium is squirted along the Earth's magnetic field to trace its movements.

But this rocket would probably just disappear into the well-mixed auroral cream. It would spend a little more than 5 min in the aurora.

I took the Nikon out of my pocket, screwed it to the tripod mount, and sighted.

"Inner stage armed?"

"Check."

"Second stage armed?"

"Check."

"Recovery armed?"

"Check."

The rocket sat there, crisp and bluish in my Nikon, until I created a fog in the viewfinder. I stood away to let the fogged glass clear.

"Experiment go?"

"Check."

"TM go?"

"Check."

"Radar go?"

"Check."

I bent back to the camera, held my breath, fingered the shutter release.

"We are approaching final count. Ten seconds. Nine. Eight. Seven. Six. Five. Four. Three. Two. One."

Abruptly the complex was bright. All the subtle blue shadows went black. A heavy roar set up a trembling in the deck. My forefinger flinched but didn't depress because the rocket remained motionless, its crawling white vapor beginning to mushroom and glow red.

I waited. The rocket moved, I pushed the button, counted to three, and released, just as the tip of the rocket approached the top edge of the frame.

I breathed. Later I would see a pathetically small pencil of light between the faint silhouettes of two winter-bare birch.

"We have liftoff," said the loudspeaker, and as the voice echoed five times among the hills, mingling with the echoing of the sound of the rocket, the intercom inside the dome came alive with cheers.

Then they did a plus count: every second to 20 s, every 10 s to 120 s, and every 20 s to 1200 s. Through it I stood on the deck, camera and hands in my pockets. The rocket contrail was caught by breezes aloft and went coiling off like an orange-and-white snake. Directly overhead, the Taurus-Orion exhaust was a sun, then a star, then a mote. Alaska resumed its winter blue. The aurora patches disappeared, replaced by long, tapered, isolated columns of light. Their lower edges were a hundred miles high, but in the illusion from my vantage they looked like white blades stuck in the hills or like searchlights beaming from the snow.

One moment on the face of a sleeping lover survives the years, as does the moment in memory of a certain bend of river or a certain face of rock. We go out to meet the moments, and we can tell the story of our going. But the story always stops just shy, and where the story stops, language fails. Like lovers and rivers and mountains, auroras are moments. They keep best if you go out to stand in the solar wind and with it blow green, white, and red.

Acknowledgments

This book, assembled and condensed from thousands of pages my father wrote prior to his death in July 2021, is a labor of love and remembrance, dedicated to the life and legacy of my beloved and wonderful father, Neal Brown. As director of the Poker Flat Research Range, my father's vision, ingenuity, and boundless curiosity helped build a place where science, adventure, and community converged. For our family, Poker Flat was not only where my father worked—it was our home (literally for one year back in 1975 with me, my sister Melody, and brother Nat), our playground, and the setting of countless stories, friendships, and lessons that shaped our lives.

My father believed deeply in making science accessible to all, whether he was speaking with fellow researchers, schoolchildren, or visiting dignitaries. He had the rare ability to communicate complex ideas with warmth, humor, and clarity—much like a Disney film that makes sense to both children and adults in different ways, yet brings everyone along for the journey. His infectious curiosity and collaborative spirit inspired everyone he worked with, and he taught us and so many others that with imagination, teamwork, and a little mischief, anything is possible.

My family and I are truly grateful to the many incredible people who were part of my father's life and work—colleagues, friends, and community members who helped build Poker Flat into what it is today. Your stories, support, and memories are part of this legacy.

A special acknowledgment and heartfelt thanks go to Sue Mitchell, our family friend, who played an instrumental role in assembling and editing my father's writings into this book, as well as to her daughter Ellen, who reviewed and catalogued all of my father's writings at the outset of this project. And a shout out must go to my father's dear friend Daniel Osborne, who worked alongside my father for so many years at Poker Flat.

But again, and in particular, credit to Sue, whose patience, dedication, and care brought these stories to life with the clarity and warmth that my father would have truly appreciated. My family and I are deeply grateful for Sue's partnership in this project.

Finally, to all those who continue to look up at the sky in wonder, just as my father encouraged: may you always remain curious, inspired, and connected to the great story of discovery unfolding above us.

Kris Brown

Contents

Chapter 1
Getting Started

Imagine that you are outside on a crisp winter night, and suddenly the entire sky is filled from horizon to horizon with rays and curtains of colored lights playing against the stars. They capture your imagination just as they have captured the imaginations of peoples of the north for eons. This collisional production of light structures in the Earth's atmosphere is perhaps the most awe-inspiring and elusive of the many energy-exchanging reactions that occur on our planet.

The aurora also captures the imaginations of scientists like me. This book is a collection of anecdotes and miscellaneous writings about how I came to spend my life involved with aurora science but especially about the major part of my working life when I was involved with Poker Flat Research Range, from 1971 to 1989. I tell the story of my involvement with Poker Flat, including some zany stories about the more than 300 major scientific sounding rocket missions and hundreds of smaller meteorological rockets launched. Along the way I talk about why Fairbanks, Alaska, is a unique place to study the aurora (and how international events shaped that reputation), how a handful of people in Alaska created their own rocket range in a few months literally out of spare parts, and how I oversaw the growth of the range into the vital and extensive science facility it is now.

My Early Life

I grew up as an only child on a wheat and pea farm about 7 miles from Pullman, Washington, and 5 miles from Moscow, Idaho. My mother graduated high school with excellent grades during the economic depression of the 1930s. She never got a chance to get a higher education because she had to work to help her family. My mom expected me to do as well in school as she had and to go to college. My dad did not graduate from high school.

N. Brown, *Northern Lights and Rocket Flights*, Springer Biographies,
https://doi.org/10.1007/978-3-032-14598-7_1

My parents were lifelong learners. They subscribed to a number of magazines—*Time, Life, Saturday Evening Post, Reader's Digest*—and talked about what they read with one another and with me. Of lasting importance to me was that they supported my many interests in reading, photography, and amateur radio.

For a few years, when I was perhaps 5–8 years old, my parents invited me to crawl in bed between them for a half-hour each night, and they told me stories. My dad had served in the peacetime US Navy from 1924 to 1928 and told wondrous tales of terrible storms at sea. He told about balls of lightning rolling down the ship's steel cables and creating a sizzling sound as they fell into the ocean. My mother told how she and her family survived the economic hard times of the Great Depression by delivering milk from a wagon drawn by a horse that knew the route without being guided. I still remember the images I created in my mind to correspond with their stories.

Until I was old enough to help my dad in our monthlong harvest, he would hire a seasonal laborer to drive our tractor. The man was usually someone who followed the harvests from Texas to Canada as the crops ripened. I loved hearing the hired men's stories about their experiences in faraway places.

I have vivid memories of my father mentoring me in how to work on a car without getting hurt. He first asked me to help him put new brakes on our family car. He warned me not to get close to the car as he jacked it up. He carefully put wood blocking under the car and then lowered the car onto it. Only then did he remove a wheel and start working to replace the brakes. Dad asked me to get him various tools and explained how he would use them. We checked or replaced brakes on our farm cars and trucks each year, and each year my dad involved me more and more. Within a couple of years, I did a brake job completely by myself, though he was close by. He encouraged me to always let someone else know when I took on a potentially dangerous job.

My dad's older brother Howdy lived just down the road and became a mentor to me. When I was a teenager, he loaned me his collection of the John Carter of Mars science fiction books. He also gave me a couple of old radios from the 1920s and encouraged me to take them apart to learn how they worked.

I got my first rocket when I was about 8 years old. It was about 12 inches long and an inch and a half in diameter. My dad and I made it out of solid wood on a wood lathe he had built. I painted it bright red. We hung it horizontally with metal hooks from a 50-foot-long wire that my dad had strung above pasture fenceposts. The rocket was powered by puncturing a CO_2 cartridge installed in the rear end. It accelerated in a blur of motion, in a whoosh that lasted a fraction of a second. The wire line the rocket rode on sang like a guitar string until the vibration stopped. The line continued to bounce up and down for a few more seconds. Launching that rocket was a thrilling experience of motion and sound.

Later in high school and college, I learned the physics and math of the action-reaction forces involved with my red CO_2 rocket. I also realized this was why the bolt nut match-head unit we made worked. My dad taught me to carefully cut the colored ignition head off a wooden match and place it between two quarter-inch bolts screwed loosely into either end of a single nut. When I dropped this assembly onto the ground bolt end first, the compression ignited the match head—and because of the action-reaction forces, the whole assembly would jump a few feet into the air. I burned the end of my fingertips many times when the match head ignited as I was trying to cut the head off. Worse still was that the burning embedded the stinky smell of sulfur into my fingertips. It usually took a couple of days before the smell faded away.

Because I was an only child growing up on a farm, my folks got me involved in Cub Scouts, Boy Scouts, and Explorer Scouts. I earned the Eagle award in Boy Scouts and the Silver award in Explorer Scouts. But what I took away from my scouting experiences was a love of the outdoors.

I graduated from Pullman High School in 1956 and went to Washington State College in Pullman to major in electrical engineering. I had great hopes that in going to college, I would do more with the future use of transistors. I failed to recognize that Washington State College educated students to work in the hydroelectric power generation and distribution industries of the Pacific Northwest. Once I figured this out, I switched over to physics, which I greatly enjoyed. My dad at first thought it was a bad choice, because he felt all I could ever do was teach other eggheads. But I met one of the most influential people in my life, physics professor Paul Bender. Paul expected that I was capable of doing anything I set my mind to. I wanted to be like him: a professional physicist with an engaging personality.

NASA Ames Research Center, 1961

Despite my enthusiasm, I had trouble with calculus and tensor math and barely got the grades to graduate college. I was lucky that NASA offered me my first professional job at its Ames Research Center, near Mountain View, California, about 35 miles south of San Francisco, right after I graduated in spring of 1961. My dad, who closely followed the US space program, told me that being a physicist was likely a good idea after all.

I was one of 50 new graduates NASA hired for each of its research centers that year in its efforts to put astronauts on the moon (and recover them safely) before the end of the decade. But first NASA had to learn how to get a spacecraft, let alone a spacecraft with a human aboard it, orbiting around the Earth at 5 miles per second and then back to the surface without burning up in the atmosphere during reentry.

After 3 days of orientation and filling out forms so I could get a secret security clearance, I was told to report to the Heat Transfer Branch close to the administration building and across the street from the library. There I met Glen Goodwin, the head of the Heat Transfer Branch, and his team. Glen was my first professional boss. He gave me several individual projects and directed me to those in his branch he thought could best help me.

We had a 3-cubic-foot steel test chamber on the ground floor. The floor was covered with metal plates that we could pick up and move. Below the floor was a steam generator from a World War II navy destroyer that had been converted to burn natural gas. It took a few hours to build up enough steam to vent through 14 stages of a steam-ejection venturi vacuum system to create the vacuum and maintain it before we were ready to use the chamber. We could open 3-foot-square doors on either side of the chamber to put an experiment inside. Opposite the vacuum port, there was a 7-inch-round domed metal plate coated with cadmium with an insulating open gap around the outer edge and a ring of cadmium mounted there. The idea was to release a high volume of nitrogen gas between the two plates and heat the gas with an electric arc of a few thousand volts and several hundred amperes of direct current. Getting the arc to start reliably was the first successful job I did. I had the machine shop in another building put a conventional car spark plug in the back side of the domed plate with the spark gap right at the surface.

Charlie, a former chief petty officer in the navy, was in charge of the steam--ejection vacuum system. He warned me that when shutting down the test chamber, I should always close the big valve between the chamber and the steam-ejection vacuum system. I never had a problem, but others who used the chamber did. As the pressure dropped, steam flowed backward into the test chamber, leaving it soaking wet. A few times I had to use a lot of rags to get the water out of the chamber because someone else screwed up.

My job was to measure the electron density of the nitrogen gas flowing out of the electrical arc and across the area where they wanted to put a model. No one had figured out how to do this over the year they had been trying. I had my second big job success in measuring the electron density by setting up a couple of microwave horns so they transmitted 10–18 GHz through the electron density area and then tuning it for cutoff frequency, which correlated with electron density. This project took me about the first 2 months of my work at NASA Ames.

My second job was to create a higher-temperature, more focused electrical arc. A predecessor had built a 2-foot coil of wire about 4-feet-long with an open hole through the middle and a high-velocity jet of compressed air running through the center. It had been sitting there for a few weeks after the other person left. Several other engineers and staff stood around me as we fired it up. It made a horrendous amount of noise, but when they connected the coil of wire to the electrical current, the wire sucked into the center, shorted out, and circuit breakers broke open.

I designed several variations on the coil of wire, but none of them worked because everything kept blowing up. My best attempt, which lasted less than one second, was when I had custom wire made out of 3/4-inch square insulated copper with a hole through the center for water to pass through to cool the coil. I remember bright flashes of electrical current and water flying everywhere.

In January of 1962 I was given a project that involved more elementary physics and tapped my electronics background. We had a monstrous device to measure electron density and atomic and molecular composition in a blow-down test chamber. NASA had written specifications for it, and a company in Palo Alto had won the bid to build it. The device stood taller than me, was 4-feet-square, and had a test area smaller than the one I first worked on. The Heat Transfer branch was instructed to test it and verify it worked before NASA paid the company that built it.

I went to the company, and right away they told me they did not think it would work because it was too big for the test chamber. NASA ultimately decided they wanted the company to be paid, so they asked me to verify that it would work in principle, even if it would not do the job the company had put in the NASA bid document.

The test area consisted of an X-ray machine pointing down across an 8-inch gap to a brass plate with angled holes arrayed around the X-ray beam center. Each angled hole led to a detector. The gist of its operation would be to have the X-ray beam be scattered to the angled circular holes with their detectors with an elaborate small computer connected to everything. I never did get it to work, but my boss Glen followed NASA instructions and paid the company anyway.

A few months after I started with NASA, I got my security clearance. My two office mates were young air force officers assigned to NASA Ames to do theoretical and mathematical work about space while earning master's degrees at nearby Stanford University. Glen decided that now I had my security clearance, the three of us should be sent over to Lockheed Missiles and Space Company in Palo Alto to a classified meeting about rockets. I expected to see something about Russian missile launches, but the meeting started with movies of liquid-fueled NASA and air force rockets blowing up before or just after leaving their launch pads. The presenter said that by analyzing our own failures, perhaps we could determine what was happening with Russian missiles at liftoff.

Lockheed and air force officials then showed us their unique space test chamber, where thin, high-temperature air moved over model spacecraft at high speed for a fraction of a second. The air was powered by an enormous discharge of electricity from a large wall of capacitors. (These were the same capacitors used in the Ballistic Missile Early Warning Systems in Alaska, Greenland, and England.) Each capacitor was about 3-feet-long, 1-foot-wide, and 2 feet in length and was connected to the rest in a power bus bar by a thin wire in an air-filled tube. The thin wire was sized such that should any one capacitor fail because of an internal short, the wire would explode and blow out the ends of the tube and shut off the electrical connection to the other capacitors. I knew the power of the electrical energy stored in just a single large capacitor and realized immediately that this was a truly ingenious design.

One day back at NASA Ames, I was notified that the navy had acquired data on a Russian missile reentry into the Earth's atmosphere taken through a periscope on a US submarine. I was pretty excited to see it, but it turned out to be only a small slip of 16-mm film used in a spectrograph with no wavelengths on it. If I had been provided a more complete readout, I might have been able to tell the difference between the part that was black-body radiation and spectral signatures of the reentry material.

As our work at Lockheed progressed, I got the chance to work on a test launch of a rocket that carried simulated hydrogen bomb warheads and masking systems. As it flew, a large chunk of carbon was ground finely and dispersed, which was designed to reflect infrared radiation. Other masking systems were employed using inflated Mylar balloons. Both could be aerodynamically slowed with this approach, enabling the combined strategies to be essentially a countermeasure to the countermeasure. A concern hinged on whether the Earth's infrared radiation would reflect the simulated warhead and the masking systems. Ultimately, 20 test flights took place with launches from, among other locations, Vandenberg Air Force Base in Santa Barbara, California, and Kwajalein in the Marshall Islands, with measurements being made from infrared telescopes in Hawaii and from rocket payloads launched from one of the Hawaiian Islands. The first Have Sled was launched without aurora, the second with aurora. The aurora reflected off the simulated warhead and masking systems at a higher altitude. A young lieutenant colonel in the air force immediately claimed he could classify the data. He could not: that had to have been done formally months in advance of a launch. But because everyone involved, including ourselves, had security classifications, they were able to keep the data secret.

Toward the end of my first year at NASA, I found a book called *Physics of the Aurora and Airglow* by Joseph Chamberlain, a synopsis of what had been learned about aurora and airglow during the International Geophysical Year of 1957–1958. At that time our knowledge of the composition, density, temperatures, and wind speeds at reentry altitudes was meager. I was taken with the idea of using spectroscopy of aurora and airglow to better understand some of what we had been experimenting with at Lockheed. It seemed to me that observations of the emissions of aurora and airglow could be used to help design materials for spacecraft that would survive reentry. I know that aurora occurs at about the same altitudes, 60 miles above the Earth, as maximum heating occurs for reentering spacecraft.

One other hint about my future involvement with the aurora concerned pigeons. I became friends with the NASA Ames staff who supported the scientists in the Heat Transfer branch. At least two of them raced pigeons. I learned that people all over the world raise racing pigeons as a hobby, which they keep in cages in their homes or on their roofs. Once a week one person goes around their neighborhood, collects the pigeons, puts them in a trailer, hauls them a few hundred miles away, and releases them at an agreed-upon time. Back home, their owners have verified trap-door timers to tell when their first pigeon gets back. As I recall, the weekly prize was only a few dollars. About once a year, this was done on a regional level, involving several states. While talking with these folks, I learned that if a geomagnetic storm occurs, they call off the release of the pigeons. Years later David Stone at the Geophysical Institute told me that birds have organic magnetic field sensors in their brains that get confused by geomagnetic storms.

Neal Brown at the Thule Air Base (now the Pituffik Space Base) in Greenland in 1963, while working for NASA

Thule, Greenland

After a year at NASA, I decided that I really wanted to get my master's degree in space physics. My good friend Ned Barmore was spending a year on a small three-man research team at the Air Force Cambridge Research Laboratory Geopole Observatory near Thule Air Force Base in Greenland. I figured I could do what he was doing, making observations of the aurora and airglow at Thule, and earn a few thousand dollars tax-free because I was not in the United States, with free room and board. I applied for and got Ned's job for the next year and resigned from NASA Ames.

At Thule, I was given status as a colonel in the air force, free room and board, and $20,000 tax-free deposited in my bank after 1 year. It was a good start toward paying for a graduate degree. I operated a 16-mm all-sky camera that took pictures of the stars and aurora, a film-based meridian spectrograph, an airglow photometer system, and a galactic-radio-receiving system sensitive to aurora, called a riometer. The all-sky camera and riometer had been built by and were operated for the Geophysical Institute at the University of Alaska in Fairbanks.

I had a ham radio license, and Geopole had a world-class amateur radio station system. Soon after I arrived, I discovered April Weather broadcasts. A user known as "April Weather" was announcing the operational countdown for the launch of rockets carrying nuclear warheads to explode in the Earth's atmosphere over Johnson Island in the Pacific Ocean. The live countdown was continuous for hours and days till liftoff, broadcast for scientists around the world to know when to compare data where they lived with the nuclear explosions. I was intrigued by the rocket countdowns.

Moving to Alaska

During my year at Thule, I enjoyed talking by ham radio with John Miller, and he encouraged me to come to the Geophysical Institute. John had graduated from the University of Alaska with a degree in electrical engineering and had worked as a Geophysical Institute employee on the first NASA satellite ground stations, Minitrack and Gilmore Creek, outside of Fairbanks. I applied for and was offered a position either in the University of Alaska Fairbanks Physics Department or at the Geophysical Institute.

My Geopole work mate Dick Rogoff argued that I should take the Physics Department position so that I could look around and decide what I wanted to do. I took Dick's advice and ended up teaching the lab section for beginning physics for nonphysics majors.

I drove my new Ford pickup with Idaho license plates to Fairbanks, arriving on September 16, 1963. Along the way, I slept under a fiberglass cover over the rear of my pickup, and at nearly every restaurant stop, I was asked if I had any Idaho potatoes for sale.

I moved into Nerland Hall dorm at the University of Alaska Fairbanks with a 40-year-old roommate named Jonathan Tong. Jonathan had been a member of Chiang Kai-shek's army but escaped from China during the Cultural Revolution. He had converted to Mormonism, and the church brought him to the University of Alaska, where he was studying accounting.

There was a cadre of Chinese students living in the rooms near Jonathan and me, including Ching Ming, who was getting his PhD under Syun-Ichi Akasofu at the Geophysical Institute, and Wilfred Wong, who got his master's degree there. Leif Owren, the head of the Physics Department, made it clear we all were to come to the Thursday-afternoon weekly seminars at the Geophysical Institute.

Neal Brown arriving in Fairbanks, Alaska, in September 1963

Bill Stringer was my office mate on the third floor of the Bunnell Building and also taught a lab. I became good friends with Bill and his wife, Sandra. Gina Ashbacker was a sophomore and Lowell Bleiler a junior in my physics for nonphysics majors lab class. Lowell grew up in a gold-mining family in Mayo and Whitehorse, Yukon Territory, and we became good friends. Despite being one of my students, Gina and I became smitten with one another and quickly fell in love.

I took a job installing fiberglass insulation for a couple of weekends to buy an engagement ring for Gina and proposed in February of 1964. We made plans to be married on May 23, 1964, at the Catholic chapel at Eielson Air Force Base where her father, Richard "Dick" Ashbacker, was a chief warrant officer in the

US Air Force. It snowed that day, and the roads to and from Eielson were icy and dangerous. My folks flew to Fairbanks for our wedding. Gina and I moved into a brand-new third-floor Hess Hall efficiency apartment, with Bill and Sandra in a similar apartment nearby on the first floor.

I really wanted to do some sort of spectroscopy of aurora and had conversations with Al Belon and Gerry Romick at the Geophysical Institute. They had a major contract using meridian-scanning photometers to capture the volume emission profile of a few prominent emissions of the aurora. They had a Hunter scanning photometer system rigged for near-infrared studies of the aurora with a liquid nitrogen-cooled S-1 photomultiplier at the Ester Dome Observatory just outside of Fairbanks. Near-infrared is just beyond the visible spectrum.

Al and Gerry had no money for the graduate work I wanted to do with the Hunter spectrometer, but they proposed it for a round of funding in the fall of 1964. In the meantime, Ben Fogle, a Geophysical Institute PhD student with a National Science Foundation grant, was looking for help with his noctilucent cloud program and hired Bill Stringer and me to work with him during the summer of 1964.

Noctilucent clouds, which are some 50 miles above the Earth, appear as silvery white well after sunset as they scatter light from the sun, far below the horizon. Ben had observational programs set up for Homer and for Tolsona and Gakona near Glennallen. Bill and I would operate ground-based cameras and photometers from Homer early in the summer. Then we would move our observational equipment to Tolsona, where we hoped to obtain photographic data on three attempts to create noctilucent clouds with water dumps from three Arcas meteorological rockets, to be launched for Dr. Willis Webb of the University of Texas at El Paso. Ben put me in charge of making sure we had reliable communications between our observation sites in the Glennallen area and the launch site at Fort Greely. The rockets would carry a canister of water to be released at altitude in an attempt to artificially create a noctilucent cloud.

After operating in Homer for a couple of weeks, Bill, Sandra, and I moved north and set up our equipment near Tolsona, about 20 miles west of Glennallen on the highway toward Anchorage. Ben had hired a high-school student to join us. I carried a 0.357 magnum pistol Gina's father bought for me on Eielson Air Force Base. At Tolsona, we were outside a small locked fenced area that surrounded a microwave system and had only a power cord and communication cable to the inside of the fence. We had been there a few days when a man drove up and said that someone had shot and wounded a sow grizzly with a cub within a few hundred yards of us. He gave us a high-powered rifle, unlocked the gate, and told us to sleep inside the fence at night. Fortunately, we never saw the bear.

Ben had set up some observing equipment and made his headquarters at the Gakona airstrip and called Bill to bring his equipment there too. I remained by myself at the Tolsona site. While I photographed several naturally occurring noctilucent clouds from Tolsona in the nights before the rocket launch, none of the Arcas rockets launched with water canister payloads successfully created a noctilucent cloud.

In the fall of 1964, Al and Gerry secured funding for me to continue taking scanning spectrophotometer data of near-infrared emissions of the aurora from the Ester Dome Observatory for what would become my master's thesis.

As I became more involved with the Geophysical Institute, I learned that T. Neil Davis, who got his PhD in auroral studies at the Geophysical Institute, was now learning how to become a sounding rocket scientist at NASA's Goddard Space Flight Center in Greenbelt, Maryland. Neil's first sounding rocket experiments were launched from a converted aircraft carrier. Each rocket carried a sophisticated magnetometer designed to measure changes in Earth's magnetic field in the upper atmosphere attributed to solar storms interacting with what is called the equatorial electrojet.

I met Neil for the first time in the spring of 1965 when he and Tom Hallinan, who also worked at the Geophysical Institute, brought their low-light-level black-and-white television camera to the Ester Dome Observatory for a test run. Neil left the Goddard Space Flight Center the summer of 1965 and returned to the Geophysical Institute as deputy director under Keith Mather.

When I defended my master's thesis on near-infrared studies of aurora, my committee recommended I continue on for my PhD, which I somewhat reluctantly agreed to do. During my defense, Al learned for the first time that I was red-green color-blind and wondered how I had aimed my instruments at the brightest parts of the aurora. In June of 1966, I earned my master's degree in geophysics. My wife Gina and my friends Sandra and Bill Stringer also graduated with master's degrees the same year. A bit over a year later, in September 1967, my first son, Kristopher, was born. That November I led the Tungsten, Northwest Territories, ground station for a Los Alamos aircraft with Neil Davis's TV flying overhead.

Then Neil asked me to help make observations of the first attempt to create an artificial aurora. So in mid-January 1969 I had my first up-close encounter with a sounding rocket and its payload in the Aerobee launch tower at Wallops Island, Virginia. The two rocket motors were already stacked, but there were problems with the payload that was being worked on. It was very cool to see the sophisticated electronics and technical machined pieces that made up the payload.

Meanwhile, in January of 1968, a B-52 crash in Greenland with nuclear weapons onboard led to Neil getting $90,000 to construct the beginnings of Poker Flat Research Range.

In May of 1968, with our second child due in October, I was having a recurring stress dream and eventually decided that I needed to pull out of the PhD program. This was a high risk move because there were no positions available at the Geophysical Institute for me, and Gina was opposed to it. But Neil created a meaningful engineering position for me, and Gina encouraged me to take it. Both she and Neil were in essence offering me a job to tide me over in the hope that after some of the stress dissipated, I would reenter the PhD program. Geophysical Institute director Keith Mather created a 3-month salaried program he called a young scientist scholar program, and I was its first recipient.

Our daughter Melody was born October 31, 1968. The summer of 1969, I built a house in Fairbanks for our growing family, and on August 25, 1970, our third child, Nathaniel, was born.

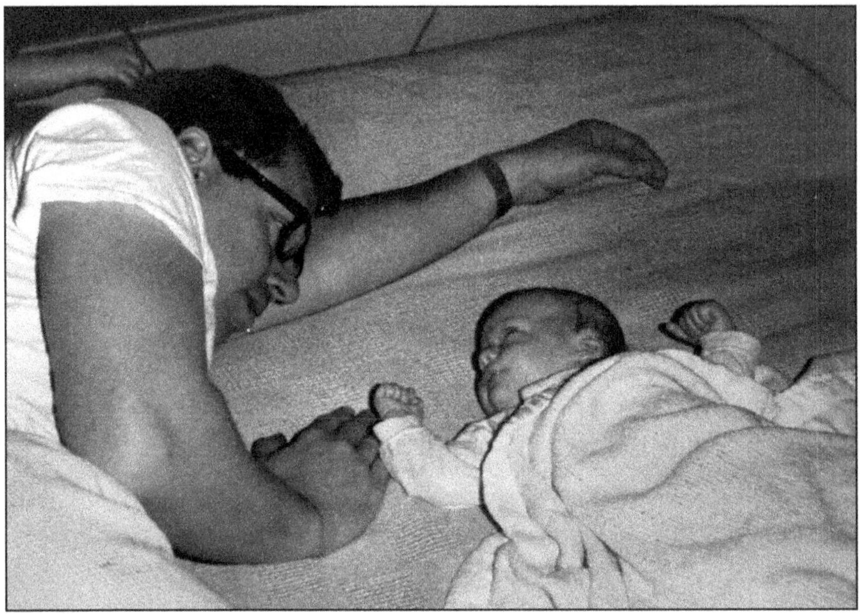

Neal Brown with his newborn daughter Melody Brown (Burkins) in December 1968

More rockets were launched from Poker Flat during the winter of 1970, while I was making observations of aurora from the Ester Dome Observatory. My focus was on the science of the aurora, and I thought it was really cool that we could now directly probe aurora with rockets, in addition to making observations from the ground and from satellites flying above the aurora. Aurora occur at altitudes too high for airplanes and balloons and too low for satellites to probe. Observations by scientists from the Ester Dome and Fort Yukon observatories were vital to making decisions about when to launch and to obtain data while the rocket was flying through the aurora.

With funding from Neil, I put together a simple public address and communication system at Ester Dome so everyone in the building would know what was happening at Poker Flat. We also came up with a plan for establishing a system to send magnetometer and riometer data and to construct a meridian-scanning photometer system for each observatory whose data could also be sent to Poker Flat over low-speed telephone lines.

In the summer of 1971, Neil was looking for someone to take charge of Poker Flat while he went on sabbatical to Norway for the fall semester. In early August he asked me to work full-time as supervisor of Poker Flat, a position I would end up holding for 18 years. I agreed on the condition that I be the only one in charge. Gina was opposed to me taking the job, and the decision may have played a role in our decision to separate in the fall of 1972. We negotiated our divorce in July of 1973 and agreed that the children would be with me for the fall semester and with her for the spring semester.

In my first real visit to Poker Flat, Neil introduced me to the crew. I told them that I was working full-time on the equipment for the Ester Dome and Fort Yukon Observatories for the rest of the year, but that I would check in periodically with them and be on call at any time. Neil was back by the time actual launches started to take place and helped me learn the ropes of launching a rocket. In early 1972, it all came together and I felt like a conductor at a concert. Everyone knew their job better than I did. I just kept everyone informed about what was needed.

In the years to come, through to my retirement from Poker Flat in 1989, I continued to provide the scientific leadership as well as the day-to-day oversight of the crew at Poker Flat as the facility expanded from one rocket launcher to five. I was pleased to experience success in all aspects of supporting cutting-edge science put together by scientists whose rocket payloads were launched from Poker Flat.

After 6 years of being a single parent for 6 months a year, in 1980 I married Fran Tannian, a former neighbor. She brought two sons to our marriage. Both were roughly the same age as my three children. Fran's two boys, Steve and Mike, straddled the ages of my three, Kris, Mel, and Nat. Initially I thought of Steve and Mike as my stepchildren, but within a few years I thought of them as my children. I have always felt truly blessed that Fran and her two sons came into my life.

Neal Brown and his wife Fran Tannian at the Tanana Valley Fair in Fairbanks, Alaska, in July 1980

After I retired from Poker Flat in June 1989, I did my first and only sabbatical, at the University of Colorado in Boulder. Although I was there to see if I could return to writing fundable proposals for noctilucent clouds, I discovered an interest in middle-school science education and teaching teachers how to do a better job with their students, who were now using computers for the first time. Back in Alaska, I created a summer science program for middle-school students and teachers that I called the Alaska Space Academy and ran it for two years in 1993 and 1994.

In 1995 I retired from the Geophysical Institute after more than 40 years and continued to work part-time supporting education outreach programs. The Alaska Legislature honored my efforts with a citation, and the University of Alaska Fairbanks honored me with an honorary doctor of laws degree in 1997. This in part led to my being asked to take over as director of a statewide program funded by NASA called Space Grant, which I ran from 2002 until retiring in 2008.

I continued to do science education outreach for local, national, and international tour groups until Fran, Molly our dog, and I left Alaska in August of 2019, settling in Lebanon, New Hampshire, near my daughter Melody and her husband Derek Burkins.

Chapter 2
Before Poker Flat

The first man-made object to orbit the Earth, Sputnik ("Fellow Traveler"), was launched on October 4, 1957, from the Baikonur rocket testing facility in the desert near Tyuratam in the Kazakh Republic in the Soviet Union. It flew west starting from 64° north latitude. Its perigee was 142 miles and apogee 588 miles. It completed an orbit around the Earth every 96 min, broadcasting beeping radio signals on 20 and 40 MHz until the transmitted signals died on October 26, 1957. The battery-powered Sputnik was 22 inches in diameter and weighed 184 pounds. It burned up on reentry from orbit on January 4, 1958.

The United States and the Soviet Union had both planned to launch Earth-orbiting satellites during the International Geophysical Year of 1957–1958. Participating countries agreed to transmit a beacon frequency of 108 MHz, and the United States was busily constructing a radio-tracking network around the world near the equator to track satellites by means of interferometry. On Monday, September 30, 1957, at an international meeting in Washington, DC, a Russian scientist informed the gathering that a launch of Sputnik was imminent and that it would transmit at 20 and 40 MHz.

About 5 p.m. on Friday, October 5, Geophysical Institute Deputy Director C. Gordon Little received a telephone call telling him that Sputnik was orbiting the Earth. Gordon was leading a radio interferometry project at the Geophysical Institute involving stars that create radio noise. Within 30 min Robert Merritt, a professor at the University of Alaska Fairbanks at the time and an expert in both power transmission and radio, and others were working feverishly to configure radio equipment to listen to and make radio-tracking measurements of Sputnik. University of Alaska student Michael Drury was chatting in the university cafeteria when Robert and Neil Davis asked for volunteers to help them put up an antenna at Ballaine Lake, just north of campus. Michael helped string up two simple east-west-aligned dipole 40 MHz antennas a half-wave apart north to south through some low treetops before

N. Brown, *Northern Lights and Rocket Flights*, Springer Biographies,
https://doi.org/10.1007/978-3-032-14598-7_2

sunset that same evening and hauled in a gigantic tape recorder. They made their first recordings by about 6:30 p.m. the same day. Within 24 h they were making accurate measurements of Sputnik crossing from west to east. They worked for 3 days without sleep.

Early on Saturday morning, October 6, 1957, Dexter Stegemeyer was sitting in his outhouse near Miller Hill Road and saw sunlight glinting off the upper-stage rocket body that put Sputnik into orbit. Dexter, known as Steg, was a neighbor of Neil Davis. He did not report the sighting until a few days later when he was chatting with Neil. As a result, the *Fairbanks Daily News-Miner* mistakenly reported that the first North American sighting was by graduate students Joe Pope and Bob Leonard and assistant professor of geophysical research John Carlo Rumi of Lecco, Italy, who saw it with their naked eyes for 5 min starting at 4:52 a.m. at 10° north of west, climbing to 60° before disappearing into sunlit sky. Skies were cloudy over most of the Western Hemisphere.

Fred Whipple of the Smithsonian Astrophysical Observatory and others would not believe the accuracy—within fractions of a second—of the radio interferometry measurements being made in Alaska. Military people in Washington, DC, were surprised to learn from Gordon Little that there was even a university in Alaska. They also questioned the measurements and assumed they were accurate only within minutes, not fractions of a second. Part of the problem was that the scientific establishment in Washington, DC, had created predictions assuming Sputnik would overfly Anchorage, not Fairbanks. The Minitrack station in Lima, Peru, got credit for the first observation on October 11. Sputnik flew right over Fairbanks many times per day, but not often over Lima. Lima had to convert their equipment from 108 to 40 MHz. The Harvard-Smithsonian Center for Astrophysics convinced the United States to build large Baker-Nunn camera systems that would provide high-accuracy tracking, if it was not cloudy, in evening and morning twilight where the satellite would reflect sunlight.

On Monday following the Friday launch of Sputnik, University of Alaska President Ernest Patty convened a faculty meeting where everyone heard the recorded signal of Sputnik.

Radio interferometry gave the time of meridian crossing. Doppler shift of the signal helped determine the orbital speed. Twinkling of the radio signal was created by electron density variations created by aurora, something Gordon Little was an expert at. This phenomenon and understanding ultimately led to the invention of the riometer by Harold Leinbach.

In November 1957, the Russians launched Sputnik II with a dog named Laika onboard. The first US satellite, Explorer I, was launched on January 31, 1958. The US Congress passed the National Aeronautics and Space Act, and President Eisenhower signed it into law on July 29, 1958. NASA started functioning on October 1, 1958, less than a year after Sputnik was launched.

Soon after equatorial-orbiting satellites were functioning, interest turned to polar-orbiting satellites. Due to the success of Robert Merritt and others with radio measurements of Sputnik, NASA contacted the Geophysical Institute to build a standard NASA Minitrack satellite receiving station at Ballaine Lake in 1960. The low operating frequency of 135 MHz meant that only low rates of data could be sent from the satellite to the ground.

In 1962, NASA again contacted the Geophysical Institute to build what is now the Gilmore Creek tracking station. It used dishes and higher frequencies, which allowed higher-speed data to get to the ground. It is nestled in the hills near 12-mile Steese Highway to avoid radio interference.

Robert Merritt foresaw the use of satellites for television and telephones in rural Alaska and proposed that the State of Alaska buy space on the Alascom Aurora satellite. Alascom had bought the military microwave communication system in Alaska and opposed state government buying space on their satellite with public funds in competition with their investment in Aurora. One of Robert's former engineering students, Fred Brown, had become an Alaska state legislator and successfully helped convince the Alaska Legislature to approve the funding despite intense lobbying by Alascom representatives.

The first effort led to establishing a telephone network with uplink and downlink from both villages with the Aurora satellite. The next step was to create Rural Alaska Television Network (RATNET) with a standing committee of delegates representing 12 community groups in rural Alaska.

Neil Davis and the First Real-Time Video of Aurora

Neil Davis was at NASA Goddard in 1963, when he learned about a television camera the Naval Research Laboratory in Washington, DC, was going to take to Fort Churchill, Canada, to attempt to record video of aurora for the first time. The camera used a 3-inch-diameter by 1-foot-long sensor called an image orthicon with 16-mm black-and-white film, viewed on a television monitor. It had 27 knobs to make it work. The naval scientists invited Neil to join them at Churchill for their first attempt. They set up near a rocket launcher, and when it got dark started fiddling with the 27 knobs. Hours later they were still changing the position of the knobs without getting decent results on the aurora overhead. While they fiddled, Neil read the instruction manual. It said after you turned it on, wait 1 h before fiddling with any of the knobs. The navy folks pooh-poohed that telling him they knew what they were doing as they had been aboard a submarine where the camera had worked perfectly. After breakup of the aurora the first night in Churchill, when the aurora was spread weakly throughout the sky, they let Neil turn it off for an hour and then turn it back on and wait another hour. Neil told me they left him alone and went to bed before sunrise. He took some video after minor fiddling with the knobs, and it turned out great. He showed them during the day, and they followed his advice that night. That was the start of Neil's first work taking video of the aurora. Within a year Neil got a major contract with Goddard Space Flight Center for a few million dollars to buy two such cameras for work in Fairbanks at the Geophysical Institute's Ester Dome Observatory. He got some grief from NASA headquarters and Goddard management, as he was not supposed to write proposals to NASA while he was in government service. (He was technically at Goddard under a grant with the National Science Foundation.) Fortunately, he was allowed to get the grant/contract with Goddard. Neil hired Tom Hallinan and bought four cameras—the whole inventory of a small company that was going out of business. No spare parts.

I first met Neil and Tom at the Ester Dome Observatory, where I was doing my near-infrared work. I helped them unpack and set up their television gear inside a 5-foot plexiglass dome. Their ability to aim in different directions to see the aurora was complicated because of the makeshift wood frame I helped them build. They were successful in acquiring good video of the aurora, so Neil wrote a proposal to use all four cameras to take video of the first-ever attempt to create an artificial aurora from a sounding rocket launched from Wallops Island, Virginia.

Neil added me to his team at the last minute, and I took part in the mission. I wound up working with Larry Sweet who, coincidentally, at the time was married to my later wife, Fran Tannian. The video Larry and I got was the best of the three cameras we operated. Ours was a side-looking shot. Neil and Tom were almost directly underneath so they were looking for something that looked like a star turning on and off. Nonetheless they were able to also get some decent video. The way-over-budget program was a huge success, even though it was supposed to take only 3 years, and we were at the 5-year point.

Operational Readiness Program

In 1961 the Soviets set off an aboveground 100-megaton hydrogen bomb on the island of Novaya Zemlya in the Russian Arctic. The United States responded by launching a number of rockets carrying atomic weapons from its Johnson Island facility in the Pacific in the summer of 1962.

Soon after, Russia and the United States agreed in the 1963 Limited Test Ban Treaty to never set off nuclear explosions in the Earth's atmosphere again.

The United States, leery of getting caught by surprise again by atmospheric nuclear weapon tests by any country, developed the Safeguard Program. This involved the Defense Atomic Support Agency, which later became the Defense Nuclear Agency and now the Defense Threat Reduction Agency, and national laboratories involved in making nuclear weapon systems such as Los Alamos in New Mexico and Lawrence Livermore in California. All of these efforts ultimately helped establish and grow Poker Flat.

Not primarily a rocket development effort, the Safeguard Program was, in essence, a US antiballistic missile defense system developed during the Cold War to protect American ICBM silos from missile attack. The program worked with companies like Thiokol to build two major rocket launcher systems: a mobile multiple rocket launcher (MRL) with a capacity rating of 7.5K (referring to the launcher's structural capacity or momentum rating in 1000-foot pounds). Thirty-two of these MRL systems were put into storage at the Sierra Army Depot, near the city of Herlong in northeastern California, until they were needed to support the United States atmospheric nuclear weapons test programs again.

MRL 7.5K launchers were a rush job but were delivered complete with interchangeable launch rails to accommodate launch lug spacing for various standard boosters, such as the Nike and Honest John. The four 1.5-inch bolts that connect the jackscrew plate to the launcher were obviously handmade by a machinist not aware of the need to spread stress from the bolt shank to the head by means of a tapered circular volume. These original bolts failed under load during low temperatures at Poker Flat in 1974 and were replaced by properly designed bolts with washers so the two plates could flex.

One of these MRL 7.5K launchers was installed at Poker Flat in the fall of 1968, on loan from the Safeguard Program. The associated paperwork made it clear that the launcher could be recalled with 24-h notice to be taken elsewhere. For the first few years, Poker Flat had a maintenance contract with the Defense Atomic Support Agency, or DASA, for it.

By the spring of 1974, Poker Flat had in place and was using three MRL 7.5K launchers and one Sandia High Altitude Diagnostic (HAD) launcher. The MRL 7.5K launcher on pad 4 failed during cold-weather rocket motor-loading operations in February of 1974.

I saw my first MRL 20K launcher at NASA Wallops in the summer of 1974 in an outdoor storage area. It was one that had been damaged during a launch at Eglin Air Force Base near Valparaiso, Florida. It had not been repaired. I saw the bent azimuth drive shaft in the pedestal. The pedestal was upright. The boom sat upright on the ground with its access door open. I remember looking at the elevation jackscrew mechanism system inside the boom and seeing angle iron welded as a V onto the circular cover plate. It connected to the elevation encoder. There was no launch rail.

Los Alamos Scientific Laboratory tasked its subordinate Sandia Laboratory to develop instrumentation to gather information on atmospheric nuclear weapon tests using aircraft and sounding rockets. Sandia maintained three of the first Boeing 707 aircraft off Boeing's production line that were already loaded with instrumentation to monitor the 1962 Johnson Island tests out of Kirtland Air Force Base, adjacent to the Albuquerque, New Mexico, commercial airport. The sensors aboard these three aircraft were continually upgraded.

As part of the earlier 1963 Limited Test Ban Treaty, the United States established a series of safeguards. One of them, Safeguard C, required the United States to maintain the capability to resume atmospheric testing in case the treaty was breached. The Department of Energy and the Defense Threat Reduction Agency continue today to maintain a facility for the potential testing of nuclear weapons at the Johnson Atoll in the Pacific Ocean.

Did I Detect an Atomic Bomb Blast in 1964?

In the fall of 1964, I was operating my electrometer-equipped spectrometer from Ester Dome Observatory, 12 miles west of Fairbanks. Just before sunrise I observed prompted emission at 760 nm, which refers to the immediate release of particles, primarily neutrons and gamma rays, during or following a nuclear event. The lithium spectral emission was coming from lithium dust created by the explosion. The data was on a piece of chart paper 6-inches-wide and 2-feet-long. I was sitting in Al Belon's office on the third floor of the Geophysical Institute when he and his office mates, Gerry Romick and Chuck Deehr, came in. They got as excited as I did.

Snow Trac Misadventure

In the fall of 1964, I decided I wanted to hunt and kill a moose to have meat for the winter. I told James Ficus, one of the Geophysical Institute machinists, that I wanted a high-powered rifle for moose hunting, and he offered me a used Winchester 300 H&H Magnum for $90. It was a beautiful rifle with a swing-over-to-the-side telescope. When I test fired it, the rifle almost knocked me down. It was a major high-powered rifle that weighed close to 12 pounds. It had a sling so you could carry it over your shoulder.

The Geophysical Institute had a Snow Trac that was kept in a locked garage at the base of Ester Dome. It was to be used by scientists and students to get back and forth to the Ester Dome Observatory if the road had not been plowed free of snow. I thought that with my farmboy experience with tractors, I could use the Snow Trac to hunt for moose on the backside of Ester Dome without any issues. I did not tell anyone what I planned to do. I told my wife Gina that I was going moose hunting the next morning, but not where, with the expectation that I would get home no later than early evening. I left our apartment dressed for a day of hunting.

I parked our pickup beside the garage at the base of Ester Dome, loaded my rifle and a bit of hunting gear into the Snow Trac, and drove it up the frozen road to the top of Ester Dome. I followed the wide Equinox Marathon trail west and down to the turnaround point about a mile and a half from the top of Ester Dome.

Before the snow came that fall, we had had several rainstorms that left a frozen layer of ice on the ground. Now, new snow lay on top of that layer. The Snow Trac was slipping a bit going down steep sections, but I thought I could easily get back up. I encountered an area of crushed snow, evidence that moose had lain down there recently. There were also wolf tracks.

A few hundred yards farther down the gently sloping valley, the track on the right-hand side of the Snow Trac came completely off as I traversed a banked part of the trail. It was late afternoon. I initially set about to put it back on by myself. The metal-connected rubber pads were much like the steel track assemblies I helped my dad refurbish on our farm tractor. But as the midOctober twilight got darker, I had made little progress.

It should not have taken much for me to realize that I had screwed up big-time, but I gave that little thought at the time and instead tried to focus on how to get back before I was missed.

After realizing I could not get the track back on by myself, I decided to walk down the north-draining creek, cross Goldstream Creek, flag down a car on Murphy Dome Road, and catch a ride back home. The sky was covered with clouds and a light snow was falling as I set out. I walked for hours without finding Goldstream Creek, the railroad tracks that ran from Nenana to Fairbanks, or the road to Murphy Dome. I heard wolves howling and was glad I was carrying my hunting rifle and pistol.

A little after midnight, I discovered that someone had abandoned a Snow Trac in much the same condition as I had. It was only when I looked inside that I realized it was the Snow Trac I had left. I must have walked in circles for hours, and fatigue and perhaps hypothermia caused me to be confused.

I sucked it up and decided that I had to walk back up the known trail to the Ester Dome Observatory. It was a long, tiring, cold slog. I learned later from my friend Bill Stringer that it was about 3000 feet of climb, in fairly deep snow. I made it to the observatory about 8 in the morning and lay down on one of the two bunk beds to rest before calling Gina.

It seemed like seconds later that I heard the door slam and Bill talking to someone on the telephone, saying he was there and there was no sign of Neal. I called out to Bill, who told me that Dan Crevensten, the ex-US Army executive officer at the Geophysical Institute, had asked the commander at Fort Wainwright for a search party to find me. Bill was supposed to wait at the observatory to be picked up by an army helicopter, since he was familiar with the area. Dan was able to call off the search before it got started. Grant La Point had been enlisted to fly his Cessna 170 over the area to search for me. He took Bill Stringer, Al Belon, and Gerry Romick along. They had spotted and followed my tracks and were amazed at my stamina.

When I got back, I apologized profusely to Gina, my thesis advisor Al Belon, and Keith Mather, the director of the Geophysical Institute.

A private rescue mission for the Snow Trac was put together that involved Eldon Thompson and Larry Sweet from the Geophysical Institute, Lowell Bleiler, and myself. About a week later, we got to the Snow Trac by taking the same trail I had used. We had the track back on within an hour. Together we continued north to cross Goldstream Creek to reach Murphy Dome Road.

We had to cut down several small trees and move them out of the way. It was about a 10-foot-steep drop down the banks into ice-covered Goldstream Creek. Eldon set cables to safely lower the Snow Trac down one bank and again to get it up the other side. The ice was strong enough to hold the weight of the Snow Trac. Before sunset we had the Snow Trac on Murphy Dome Road and were driving it back to its garage at the base of Ester Dome.

You might think I would never again let my ego overrule my common sense, but this was just one of several such events in my life. It was a humbling experience from which I learned acceptance and forgiveness. In the future I became less quick to judge others who did something they clearly should not have.

Funding for my unclassified graduate work came from the US Air Force Technical Applications Agency. Al called to tell them what I had observed and recorded. I don't remember how Al sent a copy of my data to them because we only had a fax machine. But they were alerted and got the time, date, and location where I got the data.

Several days went by before they got back to us. The lithium was man-made from a grenade exploded from a sounding rocket at 124 miles altitude over Fort Churchill, an active rocket range in Manitoba, Canada. The rocket had been launched 2 days before I saw the lithium, and upper atmospheric winds had moved it across the north at high altitude and latitude to Fairbanks.

This sparked my interest, and for the next few weeks, I aimed my system toward the east at sunrise. I saw lithium for about 3 weeks. During that time, I also saw sodium emissions at 589 nm every morning. They were left by meteors rich in sodium burning up at about the same altitude as had the lithium. Another Geophysical Institute scientist, Tunis Wentink, asked me to look for magnesium oxide at another unique wavelength in the sunrise. I never saw it, but we wrote a scientific paper that put a lower limit of magnesium in the meteors. Once deposited, the magnesium would react with atomic oxygen to create magnesium oxide.

Arctic Ice Island T-3, 1965

In September 1965, Bill Stringer was asked to take over operation and maintenance of the Geophysical Institute's chain of 16 aurora all-sky camera stations aligned along the magnetic field lines to the north magnetic pole to acquire images of the aurora borealis. Each station had a Kodak K-100 16-mm camera programmed to take one 6- to 8s exposure on Kodak Tri-X black-and-white film every minute from 6° solar depression angle after sunset to 6° solar depression angle before sunrise.

At the time, Bill and I were both pursuing our master's degrees. Bill asked if I would join him to refurbish the Geophysical Institute station on the drifting Ice Island station T-3 in the Arctic Ocean and then fly on to establish a new station at Mould Bay on Prince Patrick Island in the Canadian Arctic. Mould Bay was a year-round weather station with 12 people wintering over. The trip was estimated to take five days, at most 12 days. This sounded like a great adventure, and we could easily make up our studies for such a short time loss as this.

As we packed up our equipment to leave Fairbanks, someone told us to take lots of fresh vegetables because by the time we got there, Mould Bay would have seen their last plane with fresh food over a month ago. We superinsulated a 4- by 3-foot box and filled it with fresh green onions, lettuce, radishes, etc.

We planned to fly to Barrow (now Utqiaġvik) and then directly to T-3; 2 days later, we expected to fly on to Mould Bay and return to Fairbanks by the weekend. We flew to Point Barrow aboard a commercial flight from Fairbanks. Someone from the Naval Arctic Research Laboratory (NARL) met us in the modern terminal of the airport and directed us outside to one of their pickups equipped with huge airplane tires. When you see monster truck races with big four-wheel-drive pickups with big wheels and a camper shell on back, you have the vehicle Bill and I were escorted to. The NARL people took us to the rear of the truck, which was equipped with a small staircase and a door. There were bench seats inside and heat. We drove into the community of Point Barrow itself and then out onto the ocean beach to travel east of town to the NARL facility.

The NARL facility was spacious and well-equipped with materials, supplies, and modern tools. It looked like large Quonset hut-style airplane hangars connected to one another by sheltered tunnels called utilidors (short for utility corridors).

We spent almost a week there, with no explanation given, before we were told they were ready to take us to T-3. During our unexpected stay, we were given a tour of the facilities, which included a small zoo of wolves and polar bears. We visited the National Oceanic and Atmospheric Administration (NOAA) station. We slept in comfortable quarters, had good meals, and enjoyed the luxurious wood-lined library with its unique collection of Arctic research journals.

Ours would be the first resupply flight into T-3 since June. In the summertime the ice surface turns to slush, and puddles of water form everywhere. Each fall they have to move the buildings because they collect heat during the day and continue to melt all the time, leaving them sitting atop pillars of ice with a moat of water underneath. All other summer resupply flights had been airdrops: They rolled full fuel drums out of low-flying airplanes or just pitched stuff out that was well padded. On special occasions they used parachutes.

The plane was already heavily loaded, so we were encouraged to leave the prefab building and the insulated box of vegetables in Point Barrow. These could be brought on the next flight, which was supposed to fly back out within 24 h.

We flew the several hundred miles from Point Barrow to T-3 aboard a twin-engine RD-4, the navy version of a World War II vintage DC-3 with bigger engines. It was an uneventful trip on a beautiful day.

They had a small bulldozer on the ice island and had prepared a runway. We came in low and slow and landed easily. Bill and I helped unload the plane and watched 10 or 12 people fly away an hour or less later.

Bill and I were in awe of the nomadic human camp atop this floating slab of ice. There was an electrical generator shack with a Caterpillar diesel engine generator that ran 24 h a day. There were a dozen wanigans (small houses on sled runners instead of wheels, traditionally used as temporary shelters in lumber camps), which were connected by temporary electrical cables. We were shown to the trailer we would sleep in. It had about a half-dozen cots, but we had it all to ourselves.

We made friends with John Pierce, a graduate student from the University of Washington who was collecting and studying sea life from the bottom of the Arctic Ocean basin. There was a winch assembly in a small building that sat over a hole through T-3 (which was about 50 feet thick). John lowered a net on a cable and dragged it along the bottom as the ice island moved, and then brought it up. Small translucent critters were caught in every lift, some of which exploded after they were brought up from such great depths where the ambient pressures were so much higher. John told us that Jimmy Christensen, who was in charge of the camp, was an ex-navy officer who knew mechanics and machinery. Jimmy had a bad attitude about college kids. He did not care much about what they did as scientists and belittled them, saying that without him to fix things they would freeze or starve. He controlled the radio, etc., and he had been especially hard on our new friend John. When he was dragging for specimens, the net had snagged on something solid and snapped the suspension line. I think in the course of the summer they lost two of their three lines. Jimmy thought our friend John was irresponsible and said so under and over his breath in the dining hall. We soon learned for ourselves that Jimmy could be a problem.

There was a standing rule: no alcohol or beer on the ice island. But we did not know that when we offloaded two full cases of whiskey from the airplane. It turned out these were for Jimmy and his cronies. They got drunk and shot their rifles now and again throughout our first night. John told us Jimmy had threatened to kill him and had shot over his head at him on a couple of occasions. He hoped to get off T-3 on the next flight.

Since Bill and I were scheduled to be on T-3 only 24 h, we hurried to get the 16-mm auroral all-sky camera prepared for the upcoming winter season. We had no difficulties.

We learned how technicians with the Lamont-Doherty Earth Observatory of Columbia University took daily sightings of the sun when it was visible from their box-like shelter with its fold-away pyramid top. In the winter they took sightings of stars to locate the position and orientation of T-3, which is driven by winds. T-3 drifted most during the summertime, when the Arctic ice pack partially melted, but it still drifted at a lesser rate during the winter. More important to our auroral all-sky camera, it rotated as it drifted. The amount of drift and rotation varied daily throughout the year and was a part of the entries into our auroral all-sky camera logs.

During our first day on T-3, we learned that an engine on the RD-4 had malfunctioned on its way back to Barrow. The engine would be replaced, and we would be picked up in a couple of days. At that moment, we thought the delay of a day or 2 would give us a chance to explore T-3 a little bit now that our job was done. We ended up spending 5 weeks on T-3 before the RD-4 came back to pick us up.

The mess hall, a small prefab building about 30 by 30 feet, was the place to be any time of the day or night. The cook hailed from the Midwest. He had a family and had worked in the North many times because the money was good. He later wrote a story about our visit to T-3 that was published in the *Fairbanks Daily News-Miner*.

It didn't take me long to figure out that the crew at T-3 were much like the civilian personnel I had met in Thule, Greenland: an assortment of misfits. Many alcoholic and heavy smokers who were running away from their families. To their credit, some made good money and sent it directly home to take care of their families. Single men from failed marriages just spent it ridiculously and bragged about it. They went to warm climates and spent their money on gambling, women, and drinking.

The only communication with NARL in Barrow was via a low-frequency 165 kHz radio. Transmission and reception were both sporadic and poor. Our pickup date kept sliding day by day.

Jimmy and his trusted few drank themselves silly for the first couple of days. We rarely saw them. The generator kept running and Carl kept feeding us mounds of delicious food, but I was not comfortable with the tension between people who had summered over and the permanent crew, who had not gotten away when the first resupply flight left.

I thought I might help relieve the tension if I could get each of them to talk to their families via amateur radio. There was a completely operational radio station on T-3. The fellow who ran it was a contractor for the navy, and he communicated with his company in Huntington Beach, California. His equipment was just like what I had used in Thule, but he operated on special frequencies and for classified reasons could not let me use his radio gear. He had a hydrophone dangling below T-3 to monitor the noises they heard. Presumably he would have been able to confirm or deny whether he heard one of our submarines during a low-noise test or the noise of Soviet submarines, which were also cruising around under the Arctic icepack year-round.

Then I learned they actually had an amateur radio station on T-3, but no one was licensed to use it. The 65-foot directional beam antenna for the station was lying on the ground, but it was in excellent shape. Bill and I reinstalled the antenna in the rotator atop a nearby pole with help from a few others, and soon we were on the air. Within the next 24 h, everyone on T-3 had talked with someone back home. It seemed like the tension was less after that. We kept operating the amateur radio station throughout our 5-week stay on T-3.

There were open leads surrounding T-3, and we were warned to watch out for polar bears. John went with us during our wanderings on T-3 and carried a high-powered rifle.

A US Navy P-3 Orion with a magnetometer boom sticking out of its rear end to detect submarines flew over almost every day.

After a few days on T-3, I had established regular communication schedules with other amateur radio operators throughout Canada, Alaska, and the Lower 48. The day before we finally departed for Mould Bay, I heard a "break break" CW Morse code signal on the 20-m frequency I was using. I asked the station I was talking with to stand by. We were using voice communications, in other words a microphone and audio signal. The CW station told me he was at Mould Bay on Prince Patrick Island and they did not know anything about our coming. I quickly told him that we expected to be there sometime the next day. I tried to explain that they should have received a communication initiated by Dan Crevensten, the executive officer of the Geophysical Institute, which as I understood it had been routed through the US State Department to the Canadian Embassy in Washington, DC. He said he would tell their station chief to expect us.

The RD-4 from NARL finally arrived. We helped offload supplies and noted the presence of the boxes we needed to establish an all-sky camera station at Mould Bay. Bill and I were the only ones to depart T-3 for Mould Bay, along with the pilot and co-pilot. We told the pilots about our amateur radio communications with Mould Bay. They said, “By the way, did you know we are only going to be there four hours?” Bill and I were thunderstruck. We anticipated being at Mould Bay at least overnight. Our pilots were adamant that they did not care what we thought: they were leaving 4 h after they landed.

After several hours of flying, we were ready to land at Mould Bay around 4 a.m. It is a custom in the Arctic for aircraft to buzz the site and wait for people to come out and mark the runway in some fashion. I did not know how they were going to do that until we saw flickering flames start to appear along two parallel lines on the darkened snow-covered terrain below us. We learned later that they had thrown a little burning fuel oil into empty 55-gallon barrels that lined their runway. When they heard us coming, they drove the runway in a pickup truck, lighting a path for us.

Soon after we landed, the cargo doors swung open, and a pickup truck backed up to the plane. We quickly offloaded our boxes and were driven to the weather station. It was a well-lit single building with Quonset huts off to the sides where the 12-man crew slept and equipment was stored. We were introduced to the man in charge, who told us he had sent an inquiry off via his teletype asking who we were. This station was not a Distant Early Warning (DEW) Line site but had the same communications capability. He told us to go ahead and install our equipment and train someone to run it, but that if he did not hear from his authorities higher up, we were in trouble.

Bill and I proffered our crate of vegetables. The station leader and cook were very pleased. In fact, they woke everyone up and fed them omelets with fresh vegetables. They helped us load our prefab building onto a trailer to be towed behind their Bombardier, a motorized rig to travel in snow country that had a truck-like cab with skis for steerable front wheels. We trundled off a mile or so to a site acceptable to them and us. Bill and I were given minimal but adequate instructions on how to drive the Bombardier to the site slightly away from the main building. We quickly assembled the building. I remember being very cold as we installed the 2- by 4-inch legs of the building onto the windswept ground.

We connected our electrical power cable into a nearby metal building. This was the building from which they launched their daily hydrogen-gas-filled weather balloon. It was in constant use and heavily corroded inside. A row of ugly canisters about a foot in diameter and 4-feet tall lined one wall connected by piping. They poured water into carbon tetrachloride crystals to create hydrogen gas. On occasion it exploded, which always corroded the metal walls.

We got our station installed, Bill selected and trained an operator, a teletype message arrived sanctioning our project, and we left within the 4-h limit set by our pilots.

On our way back to Barrow, we stopped at Point Lay, one of the DEW Line sites, to pick up a load of frozen Arctic char for the animals at the NARL zoo. The fish were frozen and placed in burlap sacks that reminded me of the sacks of grain on the farm I grew up on. My guess is that NARL director Max Brewer had provided some income for the Natives in the area to provide these fish. The canvas seats along the side of the RD-4 cabin were folded up, and the entire volume was filled with

50-pound sacks of fish covered with a cargo net. Bill and I had to lie on top of these sacks of frozen fish for the next couple of hours. They were cold, but if we lay there long enough, they started to melt, and we were worried that their smell would transfer to our clothing. So we kept moving around inside the plane as we flew. We were dead meat if anything happened during takeoff or landing: no seat, no seat belt, no tie-down. An obituary about our being crushed by frozen fish passed through my mind. Talking was next to impossible because of the airplane's noise.

When we safely got back to Point Barrow, Bill and I debriefed with Max Brewer and his assistant, John Schindler. They did their best to tell us to be more understanding of Jimmy and the situation he was in. We did our best to tell them that Jimmy's anti-academic attitude toward students was a problem. They would not admit that they knew that much alcohol went out on our flight. Officially, there was to be no alcohol at all. Max and John spent an hour or 2 trying to calm us down after we expressed our fear that the situation on T-3 was explosive.

When we returned to Fairbanks, I talked to several people about our experience. Later I received a blistering letter from John chastising me for bad-mouthing NARL to others at the university. No good deed goes unpunished, I suppose.

Tungsten, Fall 1967

University of California, Los Angeles (UCLA) had a sophisticated satellite-borne magnetometer in an orbit on magnetic field lines that terminated in Tungsten, Northwest Territory, Canada. In the fall of 1967, we made coordinated observations with data from the satellite and television and magnetometer measurements of aurora in Tungsten. Tungsten was our prime ground site, but Neil Davis, Tom Hallinan, and Larry Sweet were also set up to fly over Tungsten, observing the aurora from a Los Alamos Scientific Laboratory Boeing 707 research aircraft.

Neil took me with him, and we flew from Fairbanks to Whitehorse, Yukon Territory, in a commercial aircraft. We chartered a twin-engine prop plane to fly us from Whitehorse to Tungsten, which was about 200 miles north of Watson Lake. That morning in Whitehorse clouds surrounded and covered us. I wondered if it was safe to take off and fly because of the jagged mountaintops. As we accelerated down the runway and built up speed, I wondered when the pilot was going to leave the ground. When he eventually started to climb, he pointed the nose almost straight up, and we climbed in a tight spiral until we topped out above the clouds in bright sunshine. Later Neil told me that a takeoff like that was the work of an experienced bush pilot.

In November we mounted the Tungsten campaign. University of Alaska student Lou Baim drove a flatbed truck with one of our television cubes on it from Fairbanks to Tungsten. The rest of us—the Geophysical Institute's Russ Beach, Bob Snare of UCLA, and myself—flew by commercial air to Watson Lake, where we rented a car and drove north to Tungsten. We had a small wood-frame building electrically heated in which we installed the prototype of Bob's satellite magnetometer. One night, the thermostat for the electric heater failed and overheated the building, causing the magnetometer to fail. After it was fixed, we laughed thinking how the magnetometer worked in the harsh environment of space, but we had inadvertently created an even a harsher environment to exist here on the surface of the Earth.

Night after night we collected data with the television and magnetometer. Night after night, Tom and Neil flew orbits above us in one of the Los Alamos KC-135s. The prime data were those of our ground-based magnetometer and television. The airborne TV was an ace in the hole in case of clouds obscuring our ground view of the sky.

In Tungsten, there was an open-pit hillside of nearly pure tungsten ore. Workers excavated the ore and trucked it to a large building with an enclosed crusher where it was ground to fines and slurried. The processed sandy material was carried on a conveyor belt to a magnetized roller. Here the tungsten clung to the magnetized roller a bit longer than the throw-away sand. This magnetized roller separated tungsten from debris. They were able to process ore and mine 24 h a day, 365 days a year.

We roomed in a workers' dorm and played a lot of billiards. Outside, a large lighted area had been flooded with water to create an ice hockey rink, and it was in regular use.

There were a few families living in Tungsten. One house had a television antenna mounted on its roof. We were hundreds of miles from the nearest television station and surrounded by mountains that would block any TV signal from arriving. We learned that the wife of the home with the TV antenna was completely hooked on television. During their 2 weeks off each year, she and her husband rented a room in Vancouver, British Columbia, and she watched television day and night the whole 2 weeks. She was convinced she could pick up the television signal from our experiments. We were only recording the night sky aurora on magnetic tape, but she did not understand the difference and told us she could pick out our television signal in the noise, which looked like and was called snow, the static-like pattern on a television that is not receiving any signal. We gave everyone tours of our equipment and public lectures about what we were doing.

We finished in Tungsten the week before Christmas of 1967. We packed up to drive everything back to Fairbanks. When we got to Watson Lake, we found it clouded in. Two planes a day landed at Watson Lake and flew on to Whitehorse. The morning plane could not land because of the clouds, and the chances of the clouds lifting were slim. There was only one flight a week from Whitehorse to Fairbanks, and it left from Whitehorse that evening. If we missed it, we would miss Christmas with our families. We decided to drive to Whitehorse, some 300 miles away on the ALCAN (Alaska-Canada Highway). The roads were treacherously icy. I remember arriving in Whitehorse exhausted from the drive. A four-engine plane landed as we drove into the Whitehorse airport, and we learned later that it was the one we could have flown on: it had successfully landed at Watson Lake.

Conjugate Aurora Documented

On the evening of March 11, 1967, I was aboard an instrument-laden KC-135 jet that took off from Anchorage toward the Aleutians and then turned around and started following a precisely plotted path, extending from Cold Bay northward across Alaska to Fairbanks and Fort Yukon.

At the same time, 7000 miles away, a similarly equipped KC-135 departed from Christchurch, New Zealand, and flew south toward Antarctica along an equally precise path. The two planes were part of a unique experiment performed by the Geophysical Institute to prove for the first time the theory that the aurora in the Northern Hemisphere, the aurora borealis, and the aurora in the Southern Hemisphere, the aurora Australis, were mirror images in space and time.

For a hundred years, scientists had known that both northern and a southern zones exist in which auroras frequently occur, but no one knew whether the auroras in these two regions were similar. A clue arose from studies of data taken during the International Geophysical Year 1957–1958, which showed that the northern and southern auroras had roughly similar characteristics.

Many scientists expected that some similarities should exist. The energetic particles from outer space, which strike the atmosphere and make it glow, creating an aurora, are guided along the Earth's magnetic field lines. These magnetic field lines radiate out from the Earth in a series of gigantic curves that intersect the Earth's surface at two points, called conjugate points. Since these field lines guide the particles, causing the aurora, some scientists had thought that observers stationed at two conjugate points would see identical auroras.

A simple way to see magnetic conjugacy at work is to take a bar magnet, lay it on a flat nonmetallic surface, and cover it with a piece of paper. Sprinkle metal staples or iron filings onto the paper, and they will map out the magnetic field of the magnet. The lines that lead out of one end of the magnet and circle around to the other end of the bar magnet are conjugate magnetic field lines. Auroras are analogously organized by the Earth's magnetic field.

However, most scientists studying the aurora believed that the magnetic field lines would become so distorted and weak at great distances from the Earth that they could not guide the energetic particles very well. Consequently, most theories predicted that even if auroras occurred at the same time at the two ends of a magnetic field line, they would not be identical.

Faculty and staff at the Geophysical Institute had attempted to document conjugate auroras for more than a decade before the airborne conjugate missions. They had pretty much proved auroras were conjugate with hard-won data from magnetically conjugate ground-based cameras in Macquarie Island, Antarctica, and Kotzebue, Alaska.

Professor Al Belon and director Keith Mather of the Geophysical Institute were awarded a grant to examine the question of aurora conjugacy in more detail. Neil and Al accomplished the project with a $12,000 grant from the National Science Foundation, an agreement with NASA to use their satellite, and support from the

Atomic Energy Commission (AEC). They cooperated with Dr. Neel Glass of the Los Alamos National Laboratory to equip two AEC KC-135 aircraft with all-sky cameras and photometers to measure the location and brightness of auroras.

The basic idea was to photographically document the aurora from two jets flying above the Earth, one in the north in Alaska and one in the south flying out of New Zealand. Each aircraft would fly a magnetically conjugate mission. Checkpoints and times were calculated to ensure that the aircraft would be over the points at exactly the same time. Shutter synchronizers, accurate to a hundredth of a second, were connected to the cameras to activate them in both planes at precisely the same instant.

Timing was the critical point in the planning. Flight paths over conjugate points in Alaska and south of New Zealand were determined with the aid of a high-speed computer. Because the conjugate points were in different time zones, only a few hours of darkness were available during which the auroras could be photographed. The experiment could be done only at the equinoxes (March or September), when the length of the night is the same in both hemispheres. One can only see aurora from conjugate positions for about 2 weeks each March or September, because at other times one hemisphere is in full daylight.

Flying above the clouds in these aircraft and communicating reliably by satellite helped ensure this effort's success. Scientists aboard the planes in the north and south could communicate reliably as never before through a NASA satellite. Neil played a critical role in getting NASA to allow the northern and southern aircraft to use their Advanced Technology Satellite system, which ensured that the two aircraft knew that they were in their conjugate positions at all times and that they took their photos at the exact same time.

I was a graduate student at the Geophysical Institute at the time Eldon Thompson, a GI electrical engineer, was selected to fly on the southern aircraft while I flew on the northern aircraft. I helped install a 35-mm all-sky camera in the northbound airplane at Sandia Corporation's facilities in Albuquerque, New Mexico, and then I maintained and operated the camera for the duration of the missions.

Our first step in preparing to fly these missions was to determine whether Eldon or I had any hearing problems. Eldon had been an air force crewman on F-84F fighters without any hearing protection and as a result had only marginal hearing ability remaining. As far as I knew, my hearing was fine. We drove to Eielson Air Force Base near Fairbanks and individually went into a small soundproof booth and donned headphones. It was a rather standard hearing test in which tones were changed in pitch and loudness. I toggled a switch to indicate what I heard or did not hear, and the operator wrote my reactions down. Within moments of Eldon being in the booth, the operator yanked off his headphones, saying Eldon was deaf. But somehow he stayed on the project.

To fly on these air force planes, I had to undergo training on what to do if a window or something else opened while we were flying at 35,000 feet. This was more likely than usual because the fuselage had several special openings for instruments to poke or look out of. All had been engineered to be safe, but some were unique to the plane I was going to fly on.

I joined several other GI employees involved in research aircraft missions and flew to Cannon Air Force Base in New Mexico, near the border with Texas, for training. We spent the first day being lectured to, reading, and learning about what happens to the human body when it is exposed to high-altitude air that has less oxygen in it than sea-level air, which we breathe most of the time. For example, the air in your stomach occupies a volume about the size of your fist. When you go from sea level to 35,000 feet, that air will expand because of the pressure reduction with altitude until it fills a volume almost nine times bigger. The air in your stomach will force its way out one way or the other, by your burping or passing gas.

We were shown how to don oxygen masks and told that we needed to concentrate on exhaling, not inhaling, because the oxygen was under pressure coming into our lungs through the masks. We were also told it was especially important to be aware of our rate of breathing. Too much pure oxygen would cause you to pass out. Actually, it was the displacement of carbon dioxide in your blood by the pure oxygen that would cause you to pass out. A major part of the automated breathing part of the human body gets its information on what to do based on how much carbon dioxide is in the blood.

We had two major sessions in the explosive decompression test chamber, a metal container about 8-feet tall, 8-feet wide, and about 20-feet long. The door looked like the kind of door you see in ships that leads from one compartment to another. Immediately inside this door was a small compartment, big enough for two people to sit opposite one another, and then another door similar that led into a bigger chamber with bench seats down each side. Each compartment had windows so people outside could look in and monitor what was happening.

We crawled through the two doors and sat down. An instructor told us we were going to experience over the next 30 or 40 min a slow air-pressure transition from ground level to 18,000 or 20,000 feet. He was going to give us various word, writing, and math tests and hoped we would recognize the symptoms of lack of oxygen we had learned the day before on our own—but if we did not, he would see its effects in our word, writing, and math skills. If he saw that we were being affected, he would ask us to put on our oxygen masks. Some of us would be affected more and before others.

We talked, joked, wrote, and did all that he asked of us as noisy pumps slowly pulled the air out of the chamber. The instructor told us that farting would happen but that no one would smell even "eye burner farts" because the air was being quickly pumped out or we would be wearing our oxygen masks. At something like the equivalent of 14,000 feet, he told me I was exhibiting signs of lack of oxygen. He asked if I recognized any of the symptoms and I said no. He told me to put on my oxygen mask, and I sat and waited until everyone was wearing an oxygen mask.

Eldon wrote, did calculus, and recited poetry all without showing any effect of the lack of oxygen. We hit the upper limit of the chamber, the equivalent of the air pressure at 18,000 feet above sea level, without Eldon showing any effects. The instructor then learned that Eldon had spent 3 months each of the last two summers living and working hard at a field camp atop Mount Wrangell in Alaska at 14,000 feet.

Explosive Decompression

Eldon and I were selected as the first two to experience explosive decompression. We stepped into the first small chamber, saw that the door into the second chamber was closed, and sat down on the foam cushions. We were to experience explosive decompression and then take out and plug in our oxygen masks within the first few seconds. From what we had learned the day before, we had up to a minute to plug in, which was plenty of time for such a simple task. The reason you have up to a minute is that the blood in your body is carrying a full load of oxygen at the beginning of explosive decompression. It takes more than a minute to use much of the oxygen from the blood. Indeed, when you think about it, people hold their breath or go underwater for up to a minute without suffering ill effects, for the same reason.

The entry door clanged as it was shut, set, and sealed so it would not leak. The instructors told us what was going to happen next over a loudspeaker in the chamber. They told us to look up over the door into the bigger chamber to see a flapper valve arrangement in an 8-inch hole between the chambers. It was shut. They said they had pulled the air out of the other bigger chamber and that they would open the valve over the door. When they did, our explosive decompression test would start. We were to find and don our oxygen masks right away; a couple of minutes later, they would let more air in and our test would be done.

Eldon and I sat on the vinyl-covered foam-rubber cushions on either side of the small chamber and smiled and gave each other the thumbs-up sign. We were both startled by the heavy clang of the valve between the chambers being opened without warning. Two things happened next, so fast that we later commiserated that we been stunned into inaction. All the water vapor in the air in our small chamber turned into a dense cloud that completely obscured our ability to see anything, including each other. And the air in the foam rubber seats expanded so much and so quickly that it threw us from sitting to almost standing positions. These were a couple of the tricks about this explosive decompression experience the instructors had not told us about.

We had our oxygen masks on by the time the cloud dissipated, and we waited as they let air back into our small chamber. We discussed our reaction and decided we should not spoil the experience for anyone else on our team. After we came out, we watched the next two take seats in the small chamber and then went back to a classroom to wait till everyone had experienced explosive decompression. Not one of us told the others in advance of the surprises in store for them.

We spent another 30 min with an instructor debriefing from our explosive decompression tests. He signed cards that would allow us to fly on air force aircraft that might experience explosive decompression.

Our motel was across town, and the restaurant we wanted to celebrate at was close by. The restaurant had a display case full of fresh steaks, and you selected the steak you wanted and how it would be cooked. We had visited this restaurant earlier in the week, and even though it was quite expensive, we decided that eating there would be a good reward for passing the explosive decompression test.

Eldon quietly told us that he had to go back to our motel before dinner. When we tried to argue him out of it, he revealed that during the explosive decompression test, he had soiled his pants. We knew it could have happened to us too. We went back to our motel and returned to the restaurant refreshed and ready to eat some big steaks.

At Kirtland Air Force Base in Albuquerque, we completed installation of the 35-mm all-sky cameras and prepared for a test flight over New Mexico.

KC-135 aircraft were early versions of what soon became the popular commercial Boeing 707 aircraft. The airplane I flew on was crammed with thousands of pounds of electronic equipment but had little insulation to deaden sound. Our 35-mm all-sky camera looked out of the top of the plane through a small glass dome.

I took off for the first time from Albuquerque in the late afternoon. My seat was one of three that faced backward, on a chair whose structural integrity I doubted would hold up in a crash. But the wall of heavy electronics I faced would kill me anyway in case of a crash. I had a window, which most of the other 30 scientists and technicians onboard did not.

It was impossible to converse because of the noise. We all wore headsets that clamped small speakers to our ears, shut out other sounds, and had noise-canceling microphones. We unplugged the headsets to move from station to station. The intercom boxes we plugged in to let us communicate with our own network of people and listen to other networks. I was permitted to talk on only the science network.

I decided I wanted to listen to the pilot, co-pilot, and flight engineer prior to and during takeoff. The conversation between the flight crew alarmed me right away. It was clear this plane was not mechanically perfect. I heard the flight engineer matter-of-factly tell the pilot and co-pilot that of the two electrical generators on each of the four engines, only five were working. He followed this with a long recitation of mechanical problems and aircraft ills. The crew continued to methodically check out and comment on how well different parts of the plane were working. From their conversations I got the feeling that there were several known problems that in their professional estimation were not going to keep us from taking off. The crew concluded that we were ready to take off, which left me wondering how bad it would have to be to cancel the flight. I have always assumed that nothing is risk-free—it is just a matter of how much risk you are willing to take. This time I had to remind myself that our pilots were professionals who knew how to make such judgments.

The pilots brought the aircraft to a shuddering halt at the takeoff point for the runway that ran to the southwest and started running up the engines just like they usually do on every commercial jet aircraft. They slowed the engines down and ran them up again two or three times.

They received permission from the control tower to take off, advanced power on all four engines, and the plane strained and jiggled a bit while they held it in place with the brakes. Within seconds they released the brakes and we started rolling down the runway. The runway had been built of concrete sections with a rough area between each. As we picked up speed, the time between jarring thumps grew shorter. The pilot, co-pilot, and flight engineer kept up a steady conversation about our speed and where we were on the runway.

It was a hot day, and the airport is almost a mile above sea level. These two things meant we would spend longer than usual traveling down the runway to build up enough speed, and thereby lift, to leave the ground. We reached the point of no return—the point on the runway where this airplane as it was presently configured would still have enough room to come to a safe stop—and kept on going. Soon after I heard the words "Roll 'em," and the nose of the aircraft rose slightly. A few seconds more, the thumping of the tires on the runway stopped, and I knew we were flying. I watched the end of the runway pass by only a few hundred feet below us. The land immediately off the end of the runway dropped away into the Rio Grande River valley, giving us another few hundred feet of altitude.

Suddenly the engine noise dropped dramatically, and we sank so far down it scared me. The engines were still running, but it seemed to take forever before we were again gaining altitude. I could see a half-dozen fellow passengers while this was happening, and none of them seemed the least bit worried. I later learned that I had just experienced takeoff with water-injected engines. This was one of the first three 707 aircraft Boeing had built, and they used water injection to increase the power output of the engines during takeoff. The water increased the density of the air through the engines, which resulted in more thrust. The sinking feeling I had experienced came when they ran out of the water that they had onboard for this purpose.

We flew to Beale Air Force Base in California, where we refueled for the night flight across the Pacific Ocean to Hickam Air Force Base in Hawaii. We were to fly to Eielson Air Force Base in Fairbanks, which was to be our home base. I bought a couple of fresh pineapples to share with my wife Gina when I got home, but before we took off from Hickam, I learned that our home base had been changed to Elmendorf Air Force Base in Anchorage.

We landed at Elmendorf Air Force Base in the early hours of the morning and were bused to the Captain Cook Hotel. There in the lobby, this amazingly good-looking woman with her hair piled on top of her head grabbed, kissed, and hugged me. It was Gina in a new dress, makeup, and hairstyle. She had learned of the change of home base and had flown down to be with me for a few days. She sat on the floor of our room in her skimpy nightclothes with her hair piled high and we ate fresh Hawaiian pineapple together. For the 3 or 4 days she was there we moved to a cheaper Captain Cook room to make ends meet on my per diem. Then Gina flew back to her classes in Fairbanks and I continued flying.

Our nightly run started around 11 p.m. at 33,000 feet over Kodiak Island, Alaska, and lasted about 4 h. At the same time, 7000 miles away, a similarly equipped KC-135 departed from Christchurch, New Zealand, and flew south toward Antarctica. We flew along the magnetic meridian north to Barter Island/Kaktovik and the Arctic Ocean. Then we flew back to Kodiak before returning to our base of operations at Elmendorf. Each mission night we flew across the width of the auroral oval centered roughly over Fort Yukon, Alaska, at 2 a.m., prime time for auroral activity. We flew 2 nights in a row and took a crew rest on the third night. We flew a total of ten missions over 2 weeks.

Neal Brown in September 1969 aboard a NASA Ames Convair 990

Neal's wife Virginia "Gina" eating pineapple during Neal's layover in Anchorage, Alaska

One night, as we took off from Elmendorf, the plane not only shuddered but also seemed to stagger for several tense seconds as we left the runway. I heard the control tower tell our pilot that they had seen a huge shower of sparks as we left the ground. The tail of the aircraft had hit the runway, and we might be in some danger to continue with our planned flight mission.

I felt a hand on my shoulder and saw an urgent motion for me to get out of my seat. This was unusual because we were not supposed to move until the pilot told us we had gained enough altitude. I watched as the seat I had been sitting in was quickly disconnected from the floor and thrown to one side and a door in the floor opened.

The man doing all this reached down and pulled up a tube assembly. It was a periscope used to look along the belly of the airplane. On some missions they deployed a long-wire antenna that trailed behind the aircraft as it flew. They observed the recovery of the wire with the aid of the periscope. Now they were looking for damage to the underside of the aircraft near the tail.

While all this was happening, the pilot circled the airport. When the periscope operator declared he could see no damage, the pilot announced he was going to leave the underbelly illumination lights on and we would fly close to the ground so that observers there might tell us whether they saw any damage. They did not see any damage, and as if nothing untoward had happened, they pulled in the periscope, set my seat back in place, and we gained altitude to start our mission.

Auroras were detected on all three flights on March 11, 13, and 15, 1967. During the flights, the scientists discovered they had intermittent radio communication between the two planes. This contact enabled them to adjust their flying speeds and courses more precisely to ensure that the planes flew over conjugate points exactly on schedule. After the flights I shipped the film back to the Geophysical Institute, and Eldon Thompson, based out of Christchurch, did the same. Within hours of receiving our first films, Al Belon got word back to us that we had successfully documented conjugate aurora. The photographs showed conclusively that the southern and northern aurora, at conjugate points, was identical.

The 1967 conjugate missions were successful beyond all expectations and led to missions for the next couple of years, culminating in observations using our low-light-level black-and-white television systems, which confirmed conjugacy within a fraction of a second for changes in position and intensity of weak to moderately active aurora. When activity increased beyond a certain level, the conjugate positions were modified by changing magnetic fields at the altitude of the aurora, and the aircraft were not able to keep up.

First Artificial Aurora

It was a dark and stormy night in 1969 as our small caravan of two or three cars carried the Geophysical Institute personnel across the causeway and bridges leading from the mainland to Wallops Island, Virginia, to the NASA sounding rocket launch facility. Our badges allowed us access to the island even at this late hour. We swiftly moved south past the guard shack to a tall corrugated-iron-covered launch tower complex where the Aerobee sounding rocket stood waiting.

We had arrived in the area on Christmas afternoon 1968, leaving our families and 40-below-zero temperatures in Fairbanks. We planned to set up our camera television stations to observe what we hoped would be the first-ever attempt to create an artificial pencil-like beam of aurora in the Earth's atmosphere. The program was several years behind schedule and several million dollars over budget. It seemed like this might be the last attempt to make it happen. We had been stalled out by bad weather and difficult payload problems throughout most of January. Now for the past 7 or 8 days, it seemed the major problem was the rain. The rain sleeted onto the 150-foot-tall iron structure housing the rocket. The rocket was guided by vertical rails set some 4-feet apart in a square pattern in the center of the building, leading from the floor in the basement up through the roof in the center. Once a rocket was ready to launch, blast doors would be opened around the base of the building, and doors would be swung back off the roof, allowing the rocket's clear passage as it thundered its way into space.

Now with most of the payload problems hopefully under control, the rain had been cascading down for nearly 10 days, and the roof was leaking. Water was pouring down the rails and had stalled any attempt to work safely on the rocket system, which carried high-voltage batteries and an electron gun assembly. For fear of having a problem with the water and conduction, all work had been halted.

We were a mischievous lot. We had become bored and scoured the area looking for things to do. We visited the chicken factory and the Chris-Craft boat factory, places that normally did not give tours, but when we showed up and said, "Hi, we're from Alaska," we often got a tour of buildings and facilities that weren't normally open to the traveling tourist. But now we, too, had almost run to the limit of our patience. We had to go back to Alaska. Our equipment was to be used to cover the first launches to take place at Poker Flat, starting in March. So we were getting pretty desperate. Preparations were already being made to airlift our equipment directly from Wallops Island to remote sites in Alaska, where it would be set up in February.

As we entered the corrugated-iron launch tower building, the rain was a steady rumble. There were stairways built around the working platforms, and they were made out of rough steel that allowed a good grip and were not slippery. Here and there television cameras were pointed toward the rocket system. We envisioned that, even at this late hour, there was a guard watching over the rocket from the blockhouse some thousand feet away.

If rain was going to slow things down, we were bound and determined in our enthusiasm to seal the rocket tower so that rain wouldn't further hold up the experiment. Earlier in the day, we had bought canvas, Visqueen, much rocket tape or duct tape, and some two-by-fours. We brought it into the building, and no one challenged us. So we ascended the stairs, and still no one challenged us. I remember my first glimpse of the rocket. It was shrouded in Visqueen, very unassuming, and seemed rather small for the task it was going to undertake. Perhaps not even 20 inches in diameter and standing quietly surrounded by rusty iron, almost hidden by the structure that would guide it as it left the ground. I remember looking at the part that they were working on, which was part of the payload bolted to the rocket in the tower. It had a layer of Visqueen over it, but I could see the polished and machined aluminum inside; there was obviously some intricate and sophisticated workmanship. Above it and a little bit to one side stood the remainder of the payload with its nose cone, lashed securely to a vertical member and sitting on a floor pad, also draped in Visqueen. There was a steady run of rain down the inside of the tower.

As we crawled higher and higher inside the tower, the going became more difficult. The gentle regular stairs gave way to vertical ladders. Among the eight or ten of us that were involved, we kept relaying the canvas and the two-by-fours forward to those at the top. The roof was the tricky part because we were less than 100 yards from the seashore. At ground level we were protected by a seawall that broke the wind, but at the top of the tower was a working platform. There was a trap door to go through, and then there were the door plates that could be lifted back out of the way to allow the rocket clear passage up the tower. As we got to the roof and started to lift the trap door panel to go up, it was clear that this was not going to be easy, because the wind was just screaming. The top of the tower was surrounded by a handrail, and there were blinking lights for the aircraft warning. And the top of the tower was slippery with not only water but the bird dung that had been deposited there by seagulls. We were able to lash two-by-fours between the vertical uprights on the tower, angling them down and away, and in a crude fashion build a pitched tent around the four vertical beams and angle it down to either side of the tower. This would shed the water that had been falling on the flat plate of the roof. It took several teams working short shifts exposed to the high wind and being extremely careful to finish the task, because it was so slippery and the wind was very gusty. Once completed, water still fell down the inside of the building toward the rocket and payload, although greatly diminished. We were worried that the crews would still be unwilling to work. So we brought our Visqueen to make internal aprons, again securing them by wrapping them around the four vertical bars and sloping them away, angling off the sides with string tied out in tent-like fashion. We made approximately two aprons inside the building, at the 100-foot level and about the 80-foot level, which would shed any water that fell down the center of the tower off to the side and away from the payload and the rocket. It was past midnight by the time we finished, and we felt pretty satisfied. We marveled that no one had tried to stop us, and then crawled back into our cars and drove our convoy away from the launch complex, across the causeway and past the guards, waving cheerily as we left the area and headed back to our motel to catch a morning and afternoon of sleep.

We were used to the night shift, which was when the project was to take place. The next day we started setting up as the sun was going down to prepare for a possible nighttime launch. We anxiously waited to hear the telephone ring, because we expected to be chastised for violating some of the safety rules about who can be where and when with regard to a rocket in a launching tower. Of course, the rocket had not been fueled nor had any of the igniter systems been installed. But nothing was said during the day, and we weren't wakened by a telephone call angrily accusing us of jeopardizing the mission. Later in the day, things seemed very calm. We went about our work as normal, and it seemed like everybody else was too.

Late in the day, we heard that we had greatly offended the powers that be, because we had been in an unauthorized area. It was agreed that the rocket had been potentially hazardous. The question of what to do about this renegade bunch of Alaskans was debated at an internal level at NASA Wallops, who asked NASA headquarters for guidance. We were forever grateful for the administrator at the Washington, DC, end who asked the project leaders difficult questions about what was really holding up the mission: after all, it was $2.5 million over budget and at least 3 years behind schedule, and this was its one chance to finally fly. If we hadn't jeopardized that, they said, then get on with the work and don't waste time raising hell with people. Our project leader Neil Davis may have gotten chewed out by local management, but the rest of us didn't hear a word, only that the project was rolling ahead.

Just a few days after that, the project was ready to fly. Final work was completed, and for the first time, we were in the mode of counting down the rocket for real. We had practiced countless hours. We had simulated trajectories, and we had time histories of the flight of the rocket printed out in tabular form. We knew where to point our narrow-field television cameras during each stage of the rocket flight. We had practiced stopping our cameras at selected points so that the background stars would not be smeared by the moving of the camera. We were optimizing on the one-second pulses, the long-duration, high-energy pulses that would create auroras once every 20 s or so. The electronics that controlled the electron gun beam transmitted a signal back to the ground, which was relayed from Wallops telemetry back to our site, causing a light to flash next to the 16-mm cameras that recorded what was on our TV screens attached to our television systems.

Clouds scudded in and out. We had three sites. Larry Sweet and I manned the Integrated GPS Occultation Receiver (IGOR), located in an otherwise abandoned site in a compound with chain-link fence around it. Our small University of Alaska hut was on the ground outside with its mini astrodome and television cameras. Russ Beach and Bob Bumpas manned the Eastville site about a hundred miles down the eastern shore. Neil Davis, Tom Hallinan, Gerry Romick, and others manned the Franklin City site, a little bit to the northeast. Of the three sites, Franklin City would probably be under the trajectory of the typical flight, and the electron beam would appear just as a star turning on and off, with periods dictated by the pulsing of the electron gun aboard the rocket as it traveled across the sky, a moving star amid fixed stars. The Eastville site, far to the south, would undoubtedly generate the best pictures because it was located such that it would get a full side look at the beam. It was only a question of how bright the beam was and how well they would be able to acquire a signal from it. IGOR was a good site as well: we would get long vertical pulses on our 9-by-11-degree television screen.

One evening it all came together in the premidnight hours. The weeks of practice were put to the test. Skies were clear at the three sites amid moving clouds. They would hopefully hold for as long as the 6-to-7-min flight. This was a two-stage rocket: first, a Nike booster was fired, followed by ignition of the second-stage Aerobee. An Aerobee is a liquid-fueled rocket and has a guidance package onboard to make sure it flies properly. With the heavy payload also aboard, it would not be able to achieve altitude without a small solid rocket booster. It accelerated on its way, but unfortunately the rocket started angling back straight up. Wallops Island is on the eastern shore, and its launch range extends east and southeast. This rocket started veering toward the straight up, which started to violate the rules for an errant flight. The range safety officer had removed the plastic overlays on the switches that are thrown, commanding a transmitter to send a coded pulse to a receiving package aboard the rocket that detonates a small explosive, thereby blowing up the experiment and the rocket. The technician had both hands over the switches and was thinking about firing the destruct command. Behind him stood the range safety officer. Steeped in the history of this vehicle's flight performance, with a concern that a major program would be destroyed, he made a vital decision: he reached over, grabbed the technician's hands, and held him away from the command destruct switches. Sure enough, the Aerobee slowly but surely inched over toward the east to fly a safe trajectory away from Wallops Island and the cities on the eastern shore. We saw and heard beeping on the monitors on our television screens, relayed from the rocket to the ground and from the ground station to us over telephone lines, which indicated the electron gun package was working properly. We were advised that the rocket was not flying any of the trajectories that we had anticipated; instead, it was flying high and north. Larry and I split our duties between the setting of the azimuth and the elevation track of our mount and monitoring the quality of the data we were going to acquire. We moved the mount slowly, carefully, arriving at our fixed position, standing still waiting for the one-second gun pulses. All throughout the duration of the flight, we followed the pattern we had trained: we moved, we stopped, we held, we waited. We had hoped we would see a streak, but that wasn't our prime mission. Our prime mission was to move the mount and follow the estimated trajectory and stop. Later we would look for the beam. Of course we hoped we would see it. But we did not.

The mission was over in a surprisingly short period of time: perhaps 10 min from the countdown through the last few minutes, the liftoff, the flight. Throughout that time we saw nothing. Immediately after the flight, we began to post data calibrations. The image orthicon television camera was sensitive to Earth's magnetic field, and we had to go through a series of motions, movements of the mount against the fixed stars in the sky, to calibrate what kind of distortions were introduced by the Earth's magnetic field. We realigned on the polestar as we had earlier in the evening when we first started setting up the equipment. Only after this work was completed did we drop back and review the tape that we had recorded. We moved through the videotape several times, rewound and replayed it, each time chagrined to realize that there was no immediately obvious gun pulse in it. Perhaps the mission had failed, and the gun wasn't strong enough to be seen. Then we closed down the station, it being some 2 or 3 o'clock in the morning now, and drove the hour back to the Quality Motel to the smoke-filled room of Wilmot Hess, the principal scientist behind the whole program. (Dr. Hess had by this time left Goddard Space Flight Center, NASA/GSFC, and was working at NOAA in Boulder, Colorado. But as principal investigator he was still here, and this was still his mission.) They had already reviewed the videotape from Eastville.

Russ and Bob had not seen anything and had moved their camera often and fast to different places in the sky. They rarely stopped long enough to give us a clean, good look at what they could see. Instead, their videotape was filled with slowly moving star images. It made it exceedingly difficult to pull out the gun pulse track from amid those movements.

The Eastville and Franklin City videotapes had been reviewed again and again, and apparently this process had been going on for some time before Larry and I arrived. Nothing had really been noted on the Franklin City tape. After all, one was looking for a star turning on and off, and it might be a weak star at that. It would be awhile before we had the radar track from NASA of the actual flight and could do a mathematical calculation to run the electron beam from that rocket trajectory down along Earth's magnetic field lines to the portion of the atmosphere at 68–75 miles where it could be expected to make the atmosphere fluoresce, creating the artificial aurora. All reports were negative. There were other systems there—Steve Mende had a television system also at the Franklin City site, and there were other systems around. And all the reports were pretty much negative.

Larry and I arrived late. The tapes had all been reviewed, and as a kind of strange consequence, no one bothered to look at ours. The sun was just peeping over the eastern horizon. Dr. Hess had headed north to Baltimore International Airport to catch a flight back to Boulder.

We were under tremendous pressure to get ready for the rocket-launching campaign in Alaska, so Larry and I left our videotape along with the others and headed back out to tear down our field site and pack it up so a forklift from NASA could come pick it up. The whole team was reassembled by 3 or 4 o'clock in the afternoon. Neil and I were rooming together, and when Larry and I returned that evening, Neil had a television set up in our room. He had a Polaroid camera in front of it, and he had been going through the videotapes. Every time the one-second gun pulse came on, he opened up the shutter of the Polaroid camera and allowed it to integrate several television frames. When the gun pulse was off at the end of the second, there would be approximately 30 full television pictures, and he would close the shutter.

He had a somber look on his face when we came back in. He said, "You guys really screwed up and I probably ought to fire ya." We were so tired we didn't even catch the humor in his voice, but his next words were, "You got it, you got the data, I saw it this morning the first time I put the tape on. I was able to call Dr. Hess at the Baltimore airport, paged him, got him before he got on the plane to fly west and tell him that the whole experiment was a huge success. The electron gun system really worked."

We drove to the airport that same day with literally no sleep for the last 40 or 50 hours and headed back to our families in Fairbanks.

A year later, the results of the first artificial aurora experiment were published in a major geophysical magazine with pictures of the first artificial aurora. Neil and Hess were the principal authors of the report. The pictures that Larry and I took were the only pictures in the report.

When I returned to Alaska in late January 1969, I immediately continued working to prepare the Ebert spectrophotometer system to take data on the first-ever rocket barium chemical releases that were to occur from our rocket range, Poker Flat, in the interior of Alaska. I spent February preparing the instrumentation and doing some observations on the aurora, waiting for the first launches to occur from Poker Flat in March of that year.

Barium Releases

by John Merriwether

Barium chemical releases from sounding rockets are used to track upper atmosphere winds (at altitudes above 200 km) and plasma drifts. Sunlight ionizes the barium to form ions that will drift in response to any imposed electric field. These ions fluoresce in different colors in the presence of sunlight. The upper atmosphere plasma drift and the neutral wind at altitudes above 225 km can be measured by photographing the motions of the barium ion cloud and the neutral barium cloud as they drift against the background of "fixed" stars. Triangulation from two or more camera stations establishes the successive positions of these clouds, and the ion and neutral velocities can be determined from these tracks.

Several chemical clouds are released from a single rocket payload. While confusing from this angle, the photograph is a good example of why multiple camera stations are required to understand the movement of the neutral and ion clouds. Photo by Marketa S. Murray

Spoof in Advance of the Construction of New Geophysical Institute Building

In the spring of 1968, director Keith Mather had been working hard to obtain funding from several national agencies to construct a new Geophysical Institute building on the West Ridge of the University of Alaska campus. It seemed that every few weeks, we heard a new number as to how many floors would be built and how much money would be available to build it. First we were told there would be six floors, then eight floors, and then seven floors, and the available dollars and design complexity kept changing too. In the end, only the National Science Foundation provided any funds.

I thought it might be fun to spoof what was happening. So I gathered together a number of electronic number display units and telephone stepping relays connected to a motor that rotated a drum to trigger a switch every minute. I connected these to a few 3-by-18-inch panels in a small electronic rack. Every minute, the display randomly changed the number of floors to be built or dollars to be spent. I put it just inside the entryway of the Geophysical Institute one morning before people started showing up for work and turned it on. It became a sensation, spread by word of mouth, with a parade of viewers. The best readout was the building would be 20 stories high, NSF "stock" as if on a Wall Street ticker was zero, the budget would be a low number, and number of employees would be three. After a couple of days, I took it away and dismantled it.

In June of 1970 the Geophysical Institute faculty and staff moved from our old building to the new C.T. Elvey Building on the west ridge of the UAF campus. Eventually the old building was renamed the Chapman Building.

Chapter 3
The Creation of Poker Flat Research Range

The story of how Poker Flat Research Range came to be is told much more thoroughly in T. Neil Davis's excellent book *Rockets Over Alaska: The Genesis of Poker Flat*, which he published in 2006. What follows is a brief overview from my vantage point, working alongside Neil and the others in those early days.

As early as 1964, Neil, with the help of Ben Fogle, had done a survey for NASA to find the site in Alaska best suited to launching sounding rockets. Since the 1950s, American scientists had been interested in sounding rockets to study the aurora. North America's first northern sounding rocket range was built in 1955–1956 as a joint American-Canadian venture at Fort Churchill in Canada. Although Fort Churchill was an important facility, it was a bit too far north, lying inside the auroral zone, where only breakup or postbreakup aurora occur. Still, the Canadian site was a valuable one, and since the United States, with NASA and the air force as its primary participants, had such a stake in Fort Churchill, it continued to be the United States' major center for auroral sounding rocket research throughout the 1960s.

Use of sounding rockets for research increased dramatically in the mid-1960s, when the program got a boost in the pocketbook from the Department of Defense. The early 1960s witnessed the big race between the Soviet Union and the United States in atomic weapons testing. The Defense Atomic Support Agency (DASA) went all-out in purchasing launchers, missiles, ships, and aircraft to monitor the nuclear tests, but by 1966 the signing of the Nuclear Test Ban Treaty was imminent, so DASA, in order to continue this same vein of research, considered other areas in which to concentrate. Since the barium clouds created by chemical releases from sounding rockets resemble nuclear debris, DASA saw this field as an avenue by which they could continue weapons research through the use of simulated bombs, under the guise of academic research. Surplus launchers and boosters were turned over to auroral scientists, new monies were available for research, and new launching facilities were created. Sounding rocketry had come into its own.

N. Brown, *Northern Lights and Rocket Flights*, Springer Biographies,
https://doi.org/10.1007/978-3-032-14598-7_3

Crash of U.S. Bomber Leads to Moving SECEDE III Launches to Alaska

The actual incident that led to the start of Poker Flat Research Range was something of a fluke. A major series of Department of Defense–sponsored weapons research experiments, called SECEDE III, involving barium releases from sounding rockets, was scheduled to take place in the spring of 1968 out of Thule, Greenland. This was a follow-on to a DOD barium effort that they wanted to carry out at high latitude, where the barium would be distributed into the atmosphere more.

Then, in January of 1968, a US Air Force B-52 with nuclear weapons onboard crashed near Thule Air Force Base. This was a time when "Ban the Bomb" sentiment ran high. Greenland is a protectorate of Denmark, and public indignation mixed with a good dose of political expediency mandated that American weapons experiments not be held on the soil of a Danish protectorate. This pulled the rug out from under SECEDE III.

This left DASA and the Advanced Research Projects Agency (ARPA) with six barium-release rockets and field support crews, both DOD and civilian contractors, all ready and waiting but with no place from which to launch. Fort Churchill was ruled out: after the Greenland fiasco, no one would consider anything but American soil. So under the direction of Robert Frosch of ARPA and Peter Hass of DASA, with a good deal of prodding by Col. Mike Dowe of ARPA and Neil Davis, the Alaska site survey was pulled out of mothballs, and an area north of Chatanika was selected. A minimal rocket-launching facility was to be constructed, designed specifically for this one SECEDE III mission, which was rescheduled for March 1969.

The site did not seem to have many redeeming qualities. There was no power or communication system at the site, some 30 miles north of Fairbanks. There was, indeed, nothing but a bare tract of land up the road from the abandoned gold dredges of Chatanika, near the Chatanika River and its tributary, Poker Creek. From this location was drawn the inspiration for the name, coupled with the literary inspiration of Bret Harte's short story "The Outcasts of Poker Flat," a bittersweet tale in which a gambler, a drunk, and a couple of ladies of ill repute are cast out into the cold by the self-righteous citizenry of a small Western town. Neil Davis initially named it Poker Flat as a joke, thinking that it was temporary until a more appropriate name was chosen, but the moniker stuck.

The site did, in fact, have several intrinsic characteristics that made it a natural choice: It lay on the southern boundary of the quiescent auroral zone, the best vantage point from which to study the aurora. It was in a very sparsely populated area from which rockets could be launched without interfering with any major highways or population centers. It could be integrated with the established University of Alaska, which had the highly respected Geophysical Institute, boasting excellence in all aspects of Arctic research with special distinction in auroral research. Numerous ground-based observing sites to support sounding rockets already existed throughout Alaska, and most of them were under the auspices of the Geophysical

Institute. It had easy access to a highway that lay outside the rocket flight zone but connected it with Fairbanks. Finally, the old electric power lines that had supplied the gold dredges in the 1920s were still standing and capable of supplying the present and anticipated future needs of the facility.

Like its predecessors, the SECEDE III sounding rocket project was intended to release barium. Barium is a heavy, soft metal, much like lead. It has an atomic weight of 134, which means that its outer electrons are so weakly held by the nucleus that even visible sunlight can knock them loose, leaving behind a positively charged barium ion that will lock onto the Earth's magnetic field lines. A group in Germany got the idea to release barium into a sunlit upper atmosphere while it was dark on the ground and stars could be used as tracers. They did so successfully from their launch facility somewhere in Africa. Many scientists around the world, especially those involved in auroral or magnetospheric research, wanted to use or improve the technique. The US Department of Defense immediately realized that barium could be used to simulate the effects of a nuclear weapon going off in Earth's atmosphere without the dangers of toxicity. The US atmospheric nuclear weapon tests of 1962 created several phenomena, the most notable being that the ion debris fouled up ground-based radar systems' ability to see orbiting satellites. Thus was born SECEDE I, carried out over Arecibo in Puerto Rico.

Propose, Build, Launch, Publish

To study the aurora, the process goes something like this: a scientist gets an idea, thinks it through, and writes a proposal to NASA for funding to prepare an experiment to test the idea by flying hardware aboard a sounding rocket into the aurora. The scientist has a better chance of getting funding for their experiment if they join with other scientists and propose to collaborate as a group on a problem.

After receiving the proposal, NASA sends it out for peer review to respected scientists, some of whom are likely to be friendly to the proposal and some of whom won't like it. A NASA review panel reads the results of the peer review and then compares the dollars requested by several such proposals to the money it has to fund proposals in its next funding cycle.

NASA awards the scientist money for them and their graduate students to spend the next year better defining the purpose of the experiment, and then building and optimizing the hardware to do it. NASA also assigns a team of its own, along with contract engineers, to put together a rocket motor payload that incorporates all the telemetry, radar, recovery systems, and other equipment necessary, including batteries to power it all.

The scientists whose experiments are awarded funding take their hardware to Wallops Island, Virginia, and integrate it with the other science instruments and NASA's hardware. Once integrated, everyone tests for, finds, and eliminates electronic noise problems that inevitably occur. The integrated payload is put on a large shake table that is programmed to subject the payload to the kind of vibrations it will experience during launch with the specific rocket motors chosen for the mission. NASA also runs a spin table test on the integrated payload to make sure whatever resonance it has is not sympathetic with rocket motor vibrations, which could result in failure during launch. These typically happen about 6 weeks before launch attempts. After testing and integration at Wallops, the payload and rocket motors are disassembled, then shipped to a launch range like Poker Flat, reassembled, and put onto a launcher.

The launch window, typically a 2-week moon-down period in the winter, was settled upon when the original proposal was written. Now the scientist, their graduate students, and NASA work each night as if each is the night to launch their rocket. Chances are that payload problems, weather problems, or lack of aurora on any given night will cause everyone to count down, stop, fix, or wait for days before everything comes together and they launch. As the rocket flies, NASA radar at Poker Flat tracks and NASA telemetry records what the scientific instrumentation aboard the rocket payload is encountering in the aurora. After a successful launch and flight, followed by recovery of the payload that has parachuted back to earth, everyone celebrates, packs up, and goes home.

The scientists and their graduate students will then feverishly analyze the data acquired and consult with one another while writing articles to publish their results in a major science journal, because if they don't publish, they may never get funded again.

In reality, any one scientist is doing a lot of this simultaneously. They are writing a proposal for a future launch, they are preparing instruments to fly on a funded mission, they are trying to get their rocket experiment off the ground into the conditions they want, and they are analyzing and writing documentation for publication.

In May of 1968, Neil was invited to a SECEDE I postflight meeting at Stanford Research Institute in Menlo Park, California, to review the television data. It was there that Colonel Dowe sat down next to him and negotiated not only funding to analyze the television data that Neil's team had acquired but also to build a rocket range in Alaska for the March 1969 sounding rocket launches of SECEDE III. Neil negotiated $90,000 for the construction of a minimal rocket range, which he overran by about $40,000.

Cartoon drawing by Neal Brown, jokingly showing how data from meridian scanning photometer, or MSP, instruments, and other telemetry readings from rocket launches at Poker Flat enable graduate students at the UAF Geophysical Institute to use Data General Eclipse series and other technologies to generate a lot of data, with only a small portion of the masters' theses translating into doctorate degrees

The DOD loaned the Geophysical Institute, on 24-h recall, an MRL 7.5K launcher. All the rockets and payloads would be shipped to, stored, and built up at Fort Wainwright. The DOD wanted only a warm launch structure and rocket ignition bunker at Poker Flat.

Neil secured a long-term lease from the State of Alaska on 7000 acres of land and permission from the Federal Aviation Administration and the Bureau of Land Management to use air space and to impact federal lands. Construction of Poker Flat began in mid-July 1968. Engineers Eldon Thompson, Larry Sweet, and other technical support staff of the Geophysical Institute, army and air force personnel, and technicians and engineering staff from Space Data Corporation began this task. Throughout that fall and winter, construction continued. By March 1969, the date set for SECEDE III, a rail-type launcher, covered with a movable 60-foot-tall A-frame, and a corrugated-aluminum rocket assembly building had been erected. The A-frame rode on abandoned Healy, Alaska, coal mine cars on railroad tracks so it could be pulled away from the launcher for each launch. Electric power had been brought to the range by a tie-on with the existing Golden Valley Electric Association lines. In addition to these permanent fixtures, various other storage and single-purpose tents and portable structures were brought on site.

A launch control center would be established at Fort Wainwright, and miles of army surplus WWII and Korean War field phone wire called Spiral Four were laid from the top of Pedro Dome White Alice Communication system to the rocket ignition bunker. The bunker was a small travel trailer surrounded by 55-gallon drums of frozen water to protect its occupants. (The next spring it was discovered the water freezing had cracked the drums and they were empty.)

The DOD provided balloons, helium, and operators to monitor winds as a function of altitude to generate settings so the rockets would fly where they were supposed to—and, more importantly, not land where they were not supposed to.

One of the earliest equipment acquisitions at Poker Flat in the summer of 1968 were two 40-foot-long booms hinged in the middle with a platform on one end. The other end was anchored to a 40-foot-long trailer pulled by a small tractor. They came to Poker Flat as surplus from the decommissioned Donnelly Dome super-secret Corona site, where research and weather balloons were launched to obtain data on high-altitude balloon bursts, called stratobooms, which enable the military to test satellite components in near-space conditions. Corona was one of the first satellites from the US to spy on the Soviet Union. The satellite dropped film over Hawaii, and at 40,000 feet a parachute deployed and was caught in a large net pulled behind a C-119J aircraft.

Joint Task Force Eight, a joint military and civilian organization that had overseen nuclear weapons testing, wanted a movie made of the construction, so a commercial photographer was flown up to Poker Flat. Eldon Thompson said, "We have just the thing for you." He had the cameraman lash himself to the 40-foot dual boom platform and lifted him 80-plus feet into the air. When the cameraman tried to move from one side to the other, the whole upper platform swayed back and forth about eight feet, which scared him greatly. But he got the film.

A key element of SECEDE III was that Neil's television systems would track the barium releases from Ester Dome and Fort Yukon to provide aiming data for the Ohlson Mountain Air Force Station outside of Homer, Alaska, and the Clear Air Force Auxiliary Field station radars. We were given a contract not only to build Poker Flat but also to provide the tracking system. We set out to put encoders on the azimuth and elevation shafts of the mounts that had our low-light television cameras on them. For this we set about making our own simple computer that would take this data and send it to Homer over standard telephone lines. Unfortunately, with a little more than a month to go, it was clear our homemade computer was so far behind schedule we'd never be able to use it.

The University of Alaska had an IBM 360 computer, which was overkill for the project, but IBM's contract with the university did not allow a conventional hard-wired interface with its on-campus mainframe. It would have required a lot of paperwork and weeks of waiting for IBM headquarters to tell their contractor in Fairbanks he could do the job. Fortunately, GI computational guru Bob Morton, working with GI electrical engineer Grant LaPoint, came up with a plan. The hardest part was getting data out of the IBM 360. They solved this by using the dozen or so incandescent lights on the front panel, which could be programmed to do the job we wanted. But being incandescent, they were slow to heat up, glow, and cool off. It turned out that we could break these light bulbs open and solder wires to them and not worry about heating/cooling. The next problem was easier to solve, and that was laying electromagnets on top of a keyboard that pressed on the keys to input data from Ester Dome and Fort Yukon.

Ed Gauss headed up the use of the university mainframe and was furious when he returned to discover what was happening. Fortunately, by then IBM field engineers had approved the interface effort, and top management at the university rejected Ed's complaints. We had it operational and tested it prior to the first launch and everything worked great.

I continued my near-infrared spectroscopic studies of aurora from Ester Dome Observatory and gathered data on all six SECEDE III experiments. Tom Hallinan operated one of the Ray-Scan image orthicon low-light television cameras. It was mounted on a large yoke mount in its own special dome that had been built over the landing of the stairs that led from the second floor to the roof at the Ester Dome Observatory. A small heater with a fan was used to keep the camera and its dome warm. During the countdown for one of the SECEDE III rockets, a piece of cloth got into the heater and started smoldering. Smoke filled the dome as the countdown progressed toward zero. Tom crawled up into the dome, got the smoldering burlap rag out and down onto the floor eight feet below, where Neil stamped it out with his foot. Vince Coyne, who was with the US Air Force Rome Air Development Command and had funded much of our observational program during SECEDE III, was in the room. He said in amazement, "Aren't you going to hold the count?" Tom said, "Nah, we will have the smoke out of here before the time we can take data," which was about 90 s later. Vince shook his head. Tom assured him the fan system in the dome would clear the smoke during the two minutes the rocket flew from the launch site up through the terminator into sunlight, and it did.

The SECEDE III rockets were a huge success. Radio stations throughout Alaska carried the countdown of that first launch live, and it was enthusiastically followed by the public. Carla Helfferich, the GI's public information officer at the time, wrote, "I was with friends in Hamilton Acres, standing in their yard and holding a portable radio when the countdown looked good. At first, I thought I was having hearing problems because the count was echoing oddly; then I realized that up and down Dunbar Street people were standing outside holding radios or close by open windows, and I was hearing many counts at once. When the rocket streak went up and the puff of light flared out, there were dozens of cheers echoing all around. The home team had scored a run."

One woman near Delta reported seeing the moon-sized blue barium release burst high in the sky. She said that God had opened the window from heaven but, disliking what he saw, quickly closed it.

Keeping the Range

As far as the DOD was concerned, Poker Flat had no future beyond SECEDE III. The wily and willful Neil Davis, however, had other plans. Having been asked by SECEDE experimenters and supporters to provide ground-based observations of the chemical releases from Fort Yukon, Bettles, Tok, and Ester Dome ground sites, Neil agreed on the condition that he could launch his own NASA-sponsored auroral measurements rocket from Poker Flat rather than Fort Churchill as scheduled. They agreed, and seven rockets rather than six were launched from Poker Flat in March of 1969.

The remainder of that year after the first launch season was rather grim. The DOD did pack up and move out. They rolled up 9 miles of communication cables that led from Pedro Dome to Poker Flat, which had supported the range during the March operations. Limping along on a budget of less than $500, Neil, Eldon, and Larry, in addition to their other duties at the Geophysical Institute, saw that the grounds were kept up and installed a UHF radio communications system built with surplus military equipment.

To help forestall shutting down the new facility, GI scientists hired the first two full-time range employees, launch officer Carroll Coe and wind-weighter Jim Wolff. They also put together their first very own rocket payload.

Sometime in the summer of 1969, Neil's old friend Jim Heppner called to say he would like to launch three barium-type rockets from Poker Flat in early 1970. Jim had studied at the Geophysical Institute while Neil was there and then completed his work on the data gathered in Alaska at Caltech. Neil did his master's at Caltech too. Jim headed up the section of NASA Goddard that Neil joined to learn how to do sounding rocket experiments. Jim's rocket-borne experiments were classic in their simplicity and function. His program called for near-simultaneous launch of each of his three rockets with identical rocket payloads launched from the DEW Line site on Barter Island off the Arctic coast of Alaska. Each rocket carried four barium releases. In March of 1970 from the Ester Dome Observatory, I saw a total of eight barium puffs spread over a wide range of latitude. Jim's experiment was an innovation in his attempt to understand high-altitude wind structures in the polar cap driven by solar wind interactions with the Earth's upper atmosphere.

In April of 1970, Neil and Gil Moore of Thiokol put together a rocket experiment to release an electron-grabbing sulfur hexafluoride chemical in the Earth's atmosphere that might snuff out a patch of aurora for a few minutes. Neil sent Merritt Helfferich and Bill Stringer by Snow Trac to a point in Nome Creek to observe the results. They were only supposed to be there long enough to capture pictures of the release and return to Poker Flat within 12 h, but bad weather delayed the launch. The two had no sleeping bags, not much food, and no radio communication. They had trouble keeping the heater and Snow Trac running because the fuel tank had rust in it. Neil took his personal snowmachine out and brought them gasoline and other supplies. Then Hans Nielsen made a supply drop from his personal airplane, including a sheet of paper with a fancy ribbon, a wax seal, and "official" signatures awarding Bill and Merritt "Cheechako of the Year" for going out without the proper winter gear (in Alaska, a *cheechako* is a person new to the region who has not yet survived a winter). As an additional dig, the crew at Poker Flat wrote a message on the nosecone of the payload that read "To Bill & Merritt, With Love, From Neil, Tommy and the Boys" and included a photo of it with the airdrop.

More Launches

By the summer of 1970, it was obvious that the facilities at Poker Flat were inadequate. Other NASA-supported sounding rocket experimenters contacted Neil about launching during the winter of 1970–1971, and Neil and fellow University of Alaska scientist Gene Wescott received funding from the National Science Foundation (NSF) for a small rocket-borne barium-release program. Neil had also written a proposal to NSF for funding to establish a sounding rocket radiotelemetry station at Poker Flat. Thirteen sounding rockets were scheduled to be launched between the upcoming November and March, so construction began to expand the range and its capabilities. A second launcher was installed, and for the first time, permanent telemetry and radar stations were established. The range's capacity was vastly improved by the long-term loan of a portable Sandia UHF telemetry station and the NASA VERLORT (Very Long Range Tracking Radar) station, which had been used in the early days of manned space flight and was capable of tracking a beacon-equipped rocket or spacecraft to the moon.

NSF funded the construction of a 40-by-40-foot blockhouse to house most of the range's major functions: range and payload control, telemetry, and wind-weighting. Sandia Corporation (now Sandia National Laboratories) donated a completely instrumented set of telemetry equipment to Poker Flat, and the blockhouse was in operation with the Sandia telemetry inside by that fall. Other improvements included new roads, better on-site communications, the installation of a microwave link to Pedro Dome for both telephones and scientific data, and the addition of several wooden storage and work buildings.

Neil installed a three-component magnetometer system near the blockhouse, which helped people on range see what the aurora was doing.

NASA Wallops sent two radar system trailers on railway cars to Poker Flat. One contained a VERLORT system and the other an MPS-19 radar system. When John Benevento, the NASA Wallops radar expert assigned to the project, opened the door to the MPS-19, he saw circuit boards strewn on the floor and electronics hanging by wires. Fortunately, he was able to piece it all back together. Larry Sweet built a concrete pedestal platform for the VERLORT, which was in use for more than a decade after.

In November of 1970, Larry Cahill from the University of Minnesota launched two experiments aboard the NASA workhorse two-stage Nike-Tomahawk rocket systems, and Neil and Gene launched two barium-release experiments. Cahill's experiments were successful, but the Bullpup-Apache rockets Neil and Gene were using did not get high enough into the Earth's atmosphere for their experiments to work properly.

In February of 1971, Hugh Anderson, Paul Cloutier, and a gang of their graduate students from Rice University brought a densely packed set of instrumentation to Poker Flat for launch aboard a NikeTomahawk.

How to Keep a Sounding Rocket from Crashing Back to Ground

A sounding rocket goes from the ground to near-Earth space pointy end up. Once the payload separates from the rocket and is free of the Earth's atmosphere, the pointy end stays pointing up through apogee and descent until it encounters the atmosphere again. Then things can get crazy; the rocket spinning around its own axis and tumbling end over end, unless you make some design changes and add a parachute recovery system.

First you have to figure out the center of gravity for the payload. You do this by laying the payload lengthwise across the equivalent of a knife edge and moving it back and forth until it balances. To change this center of gravity, you move things around inside the payload. If you succeed in moving the center of gravity to the right spot, now as the payload descends into the atmosphere again, it will tend to lie flat, like a falling pencil.

In reality this happens about 60 miles above the Earth's surface. The payload could end up as much as 30 miles from the reentry point. Because the density of the Earth's atmosphere increases exponentially, this typically happens when we can no longer see the payload or track it with radar or telemetry. We call this loss of signal, or LOS.

Now we can wait until it gets down to about 25 miles altitude before we deploy a small, rugged parachute called a drogue chute, often smaller in diameter than the payload. By 4.5 miles, the sharp cutters around the bag that holds the main chute explosively cut it open so the drogue chute can now pull the main chute out. That orients the payload so the pointy end is down. Deploying the drogue chute also turns on a radio locator beacon that can last for several days even at 40° below zero.

For our studies of the aurora, all this typically happened at night. We would fly out the next day in a small single-engine airplane with the best radar and telemetry data we had prior to loss of signal. The radio beacon typically led us to see the payload and parachute together on the ground within an hour of flight.

Soon after I left Poker Flat in 1989, they started to put small radios inside the payloads with antennas on the outside that transmitted to Earth-orbiting satellites. They typically could get a payload's GPS coordinates within half a day. Later still they added a radio unit to each of the rocket motors as well as for things that were purposely ejected like the nose cone.

In March Karl Theobald of Sandia launched two of his Caribou V experiments involving measurements of the ultraviolet emissions of the aurora, the first aboard a Terrier-Tomahawk rocket motor system. This was Poker Flat's first mission failure: booster instability caused the payload, along with the second-stage motor, to be pitched off shortly after liftoff. The second Caribou V experiment was successful and was also the first successful recovery of a parachuted sounding rocket payload at Poker Flat.

Also in March that year, Neil launched two Nike-Tomahawk mother-daughter experiment payloads he called HAWKS with Alan Johnston as his graduate student.

Wallace "Wally" Murcray launched two Nike-Tomahawk payloads that contained photometers filtered to observe specific emissions of the aurora to determine the brightness of the emissions as a function of altitude.

Hugh Chivers of the University of California launched two Sandia-sponsored payloads, which exposed films to the radiation in the aurora at altitude and were identical to ones the astronauts had exposed on the moon and brought back to Earth. The rocket-borne films were sealed in compartments prior to recovery. The hope that elements in the solar wind captured on the moon would appear in the films launched from Poker Flat did not materialize, perhaps because the films spent so little time exposed during the sounding rocket's flight, in contrast to the lunar exposures.

The Ester Dome Observatory

Ester Dome was the major optical observatory of the Geophysical Institute and was used extensively for the first SECEDE III rocket launches from Poker Flat. During the winter of 1966–1967, my wife Gina and I lived in a small home just below the

Ester Dome Observatory. The electrical currents associated with the aurora, even when far away, create enough of a change in the Earth's magnetic field to create a current in electrically conductive rock. I rigged one of Vic Hessler's Earth current devices at Ester Dome to turn on a light in our bedroom to give me an early warning of aurora. In addition to making my own observations that winter, I also maintained and operated the instruments that were making routine measurements at Ester Dome. On the roof under 3-foot-diameter Plexiglas domes were several instruments. One was a 16-mm film all-sky camera system that was used to take one 8-s-long exposure each minute from evening dusk until morning twilight. Under another was a four-barrel telescope pointed into the magnetic zenith. Under another was a simple spectrograph that documented the spectrum of overhead aurora. In addition to these, there was a meridian spectrograph that used film in its own hut attached to the building. A 30 MHz antenna system off to one side was a part of the riometer. Al Belon and Gerry Romick had also built and installed meridian-scanning photometer systems at Ester Dome, Fort Yukon, and Kaktovik on the Arctic coast.

In addition to the regular commercial telephone in the building, there was another telephone connected only to a similar telephone on a desk at Poker Flat. I don't recall exactly how we knew where we were in the countdown, but I think some sort of simple public address system had been installed in the building. I told Neil that the Ester Dome Observatory needed an intercom and a public address system connected to Poker Flat. Neil agreed and tasked me to put together and install a simple system, which worked very well during the second launch winter of 1969–1970.

I continued making my own near-infrared measurements at Ester Dome during the winters of 1969–1970 and 1970–1971. The scientists whose rockets were being launched at Poker Flat typically were at the Ester Dome Observatory when they made their decision to launch. I often talked with them and Neil about the things they wanted to know before they decided when to launch their rocket and what scientific data they wanted afterward from us. It was from these conversations that I suggested to Neil that I rebuild the meridian-scanning photometer system so that the data from Fort Yukon could be read out at Ester Dome or anywhere else. Neil agreed, and I designed and started building the system in the spring of 1971.

In early 1971, Neil received roughly $125,000 in funding from NSF to improve launch-decision science at Poker Flat. The scientists launching their rockets from Poker Flat could locate themselves at Poker near their rocket and support crew or communicate by telephone from the Ester Dome Observatory or the Fort Yukon Observatory north of Poker Flat. Neil and the Poker Flat crew had installed a sophisticated three-axis magnetometer a few yards from the blockhouse and built a very small observing hut alongside the top of the blockhouse called the birdhouse. The birdhouse allowed an unobstructed view of the aurora and was connected by intercom to the launch control center inside the blockhouse. The wood-frame birdhouse with its glass roof was less than 1000 feet from the launchers, and officially no one was supposed to stay in it during an actual launch.

The single telephone line from Ester Dome or Fort Yukon to Poker Flat was proving to be inadequate when several supporting scientists or graduate students were at these observatories. After the end of the spring 1971 rocket launch season, Neil and I agreed that we needed some simple intercoms at these observatories and a small countdown amplifier speaker system so everyone in the observatory knew the status of the rocket countdown. Neil underwrote the relatively minor costs involved, and I put together and installed equipment at the Ester Dome Observatory.

I continued to talk with Neil about how we might get more prelaunch science from Ester Dome and Fort Yukon to scientists who elected to make launch decisions from Poker Flat. By early summer of 1971, Neil funded me to put together a simplified meridian-scanning photometer system at Ester Dome and Fort Yukon and a simplified data transmission system that would send data on the geographic position of two colors of the aurora and magnetometer readings from Fort Yukon and Ester Dome directly to Poker Flat, where they would be displayed on extra channels of the chart recorder in the blockhouse that displayed data from the Poker Flat magnetometer system.

Al Belon and Gerry Romick gave me complete access to their meridian-scanning photometer systems to adapt them for this project. I chose many simplifications as I was designing and fabricating the equipment involved. I had an amazing bit of serendipity when Techtronic asked me to submit a letter saying why I wanted two of their $25,000 memory storage oscilloscopes, one for each observatory, and then donated them to my project. I hired Charlie "Soup" Campbell from the Geophysical Institute electronic shop to help put everything together and test it and then sent Soup to install the equipment at Ester Dome.

Poker Flat Director Neal Brown in 1977

Chapter 4
First Full-Time Manager of Poker Flat

During the summer of 1971, the range had 18 full-time employees, most of them involved in construction and scientific instrumentation. Everyone knew Neil was going on sabbatical to Norway that fall and that Larry Sweet was leaving to go back to graduate school. We all wondered what Neil would do about Poker Flat in his absence. While Eldon Thompson could do almost anything, he needed some guidance, and in the past, Larry had successfully interpreted for him what Neil wanted.

Getting the Job

It was midafternoon in early August 1971 when Neil Davis called me into his office and asked if I would be interested in becoming the first full-time manager of Poker Flat. I told Neil that while I was very interested, I wanted a day to think about it and discuss it with my wife, Gina. Neil was glad I was interested and thought it was a great opportunity for me and the Geophysical Institute. I had a hard time keeping my enthusiasm in check as I left Neil's office.

Before I took on the job, no one had worked full-time managing Poker Flat. Neil, Larry, and Eldon put in untold hours, but each had also carried other work-related responsibilities as well. Neil would be leaving soon for Norway. Larry, Neil's right-hand man at the GI with regard to Poker Flat, was on unpaid leave to complete a master's degree in engineering management. Frankly, it was not clear how Poker would survive the absence of Neil and Larry at the same time.

I sought out Gerry Romick and told him Neil had asked me to step in to manage Poker Flat. Gerry and I had talked many times throughout the summer about the possibility that I would be offered the job and what I might do with it.

N. Brown, *Northern Lights and Rocket Flights*, Springer Biographies,
https://doi.org/10.1007/978-3-032-14598-7_4

I had also spent time talking with others about the opportunities and difficulties of running Poker Flat. Most thought Neil, Larry, Eldon, and the range crew were suffering burnout. Many thought communication was at a low point. Neil had too many irons in the fire. Larry seemed frustrated when he couldn't bring about the kind of orderliness he seemed to like. He told me that he was not interested in being a part of Poker Flat's future.

Larry and I had known each other and worked together for several years, and we built homes within a quarter-mile of one another in the Musk Ox subdivision, in the hills north of Fairbanks. Larry's wife, Fran, and my wife, Gina, were good friends. They were both pregnant with their youngest boys and laughed a lot during late-night telephone calls the winter of 1969–1970, while Larry was at Poker Flat and I was at Ester Dome for the SECEDE III operations. They still joke that their sons Michael Sweet and Nat Brown were friends before they were born.

Eldon Thompson was a bit of a wild man. We all knew that Eldon could get any job done, no matter how little he was given. Yet even though Eldon seemed to be in his element building Poker Flat out of almost nothing, he also seemed frustrated at not having enough money or people to get the job done the way he wanted to do it. I decided that communication was a problem. Scientists and staff felt that Neil had not communicated well enough with them, and Larry and Eldon and the range crew were not communicating well with each other. I could sympathize with the individual complainers, but communicating with Neil was not a problem I had experienced personally. He and I had roomed together for 4 weeks during the first rocket-borne artificial aurora creation experiment back in January of 1969. Neil and his wife, Rosemarie, were gracious hosts, inviting students and staff to their home for impromptu parties, and they came to our graduate student parties as well.

Gina and I had discussed my taking the job several times. She thought it was a mistake. She thought I should stay on the academic and research path I was on. So when I came home that evening, there wasn't much more to say. She knew I wanted the job.

The next day I met again with Neil and told him some of my concerns with regard to past communication problems. I did not give any specifics, and Neil did not seem to be offended. I told him I really wanted the job, but that I would take it only if I got full authority and responsibility right from the start. I did not want to be just a front man carrying out orders. I told Neil I would take the job for a minimum of 1 year and was considering a maximum of 4, lest I lose touch with the research I was doing. Neil knew that being a scientist and doing cutting-edge research were important to me. We both knew that managing Poker Flat was primarily administrative, yet we also knew if it were not managed by someone really interested in science, it would quickly fail. Neil expressed how glad he was with the attitude I had taken and said he would support me. I had a major project underway for Poker Flat in the Geophysical Institute electronic shop to build and install meridian-scanning photometers at Ester Dome and Fort Yukon. I told Neil I wanted to see if Eldon could keep things going at Poker Flat until I completed that project in time for the January launch season. Neil, concerned about the communication problems I had raised, told me he thought I should take charge of the crew at Poker Flat right away. I agreed.

My First Day at Poker Flat

On August 18, 1971, the day his memo appointing me supervisor of Poker Flat was distributed, Neil took me on the 30-mile drive northeast of Fairbanks to Poker and introduced me as its first full-time supervisor. I told the assembled crew that I was deeply involved in getting new equipment assembled that would bring prelaunch scientific data to Poker Flat to help the scientists make their launch decisions. I told them frankly I had no prior experience with rockets themselves but that I had been involved in making observations from Ester Dome of the first SECEDE III launches from Poker Flat. I also shared with the crew that I cared a lot about the science that could be done from Poker Flat and that I hoped the self-reliant experiences I had had growing up on a farm would make it easier for me to help them get their jobs done. I promised to be available anytime, anyplace, then said that from now until late December, I would be spending almost all of my time in the Geophysical Institute electronic shop.

I may have seen a rocket launcher at Poker Flat before, but this time I took a better look at the two launchers. Each was mounted on a large, reinforced-concrete pad about 25 feet square and eight feet thick buried in the ground, on top of which was installed a 6-foot-diameter round steel structure. About 8 feet up the vertical structure was a horizontal bearing, and the launcher was a long cantilevered section that hung from bearings at the base of a sort of steel dunce cap that sat above the horizontal bearing. On top of the dunce cap structure, an electric motor-driven jackscrew connected it to a point about one-fourth of the way along the launch rail system. This jackscrew was used to bring the launch rail from the horizontal to the near-vertical position for launch.

Neil and Eldon proudly showed me the new clamshell protective heating system Eldon had the crew installing on pad 1 and the nearly complete installation of a High Altitude Diagnostics launcher donated to Poker Flat by Sandia Corporation not far from it.

The rocket motor/payload-heating clamshells Eldon was constructing on each launcher consisted of a steel frame box attached to the launcher. The launcher itself would be the roof. Along either side of the launch rail hung an L-shaped frame, each forming one-half of the final box that ran the length of the launcher. The frame was constructed of lightweight Unistrut steel that in cross section looked like the letter *C*. The frame was built so that lightweight sheets of 1-inch-thick, 4-by-8-foot sheets of Styrofoam could be slid into the cross section of the Unistrut. Large sheets of Visqueen would be wrapped around the outside to help protect the Styrofoam from the wind. Each clamshell box frame would be approximately 8-feet square by the length of the launcher. The L-shaped sections were each attached to long rotatable pipes that ran the length of each launcher. An electric motor and chains installed on the top of each launcher would be used to rotate the clamshells open. With the launcher in the horizontal position, the clamshells could be partially opened so crews could walk under the launcher and install the rocket motors and their payloads. If the clamshell failed to open, the rockets could still be launched. In the process the replaceable Styrofoam and Visqueen would be destroyed, which would be a relatively minor loss. As it turned out, they were destroyed by the launch blasts anyway, but they did their job of keeping heat around the rocket and payload before launch.

A hundred or so feet from each of these launchpads stood a wood-frame heater hut building on its own concrete pad. Each was equipped with a 100-kW electric heater. A long wooden aboveground structure about 6 feet high and 4 feet wide ran from each heater hut to a launchpad. Each of these structures held two pieces of 24-inch Davidson Ditch water pipe stacked one above the other to take heat from the building to the rocket on the launchpad. Vermiculite insulating materials filled the wood boxes surrounding these pipes.

Neil and Eldon also gave me a tour of the newly constructed concrete blockhouse. I saw three telephones on the launch controller's metal office desk. One of them was ringing, but I couldn't figure out which one. I asked Neil how he knew which telephone to pick up, and he said, "When in doubt, just put your hands on the telephones; you can feel which one is ringing." He pointed out that one of telephones was a dial telephone and added proudly that it was the only telephone at Poker Flat connected to the university telephone system. He pointed to the other two telephones that had hand cranks on them, and said one of them was the Poker Flat hot line to the Federal Aviation Agency (now the Federal Aviation Administration, or FAA) and was used to coordinate rocket countdowns with local air traffic. I don't remember what the other hand-crank telephone was used for.

Overhead there were three colored light bulbs in each room of the blockhouse. I asked Neil what the colored lights meant, and he said if the green one was lit, everyone in the blockhouse was relaxed but counting down a rocket. If the yellow one was lit, a launch could happen in the next 10 or 15 min, and no one was to leave their blockhouse station. If the red one was lit, then the final few minutes of a countdown was in progress. I told him I was color-blind enough that I probably would not be able to tell the difference between the yellow and red bulbs.

Poker Launch Buttons

It's pretty thrilling to know that a simple button used to launch a model rocket is very much the same as the button used to launch a sounding rocket.

During my first launch season in the spring of 1972, we had intercoms connecting everyone. The Rice University scientists and graduate students at Ester Dome and Fort Yukon spoke at length while the rest of the range crews listened in. They spent a lot of time discussing what they wanted, as if they had not decided that the year before when they built the payload. We heard them express regrets for events that would have met their criteria but they didn't launch. It got so hilarious that I bought a laugh button and broadcast the sound over the intercom, much to the enjoyment of everyone involved.

When we shut down Ester Dome observatory and built Poker Optics where the launch scientists could now watch the skies, I put a fake launch button next to the intercom. Everyone knew it did not work, but this time it was the scientist who pushed the button when they screwed up. Again, everyone got a kick out of it.

Despite working hard to make sure that everyone involved knew the criteria for launch, the scientists often changed the criteria at the last minute just enough to throw a bit of confusion into the mix. This led me to give a crystal ball to the launch controller. Several times each night, she would be asked if the clouds were going to disappear or if the aurora would come out strong enough to launch.

After I retired from Poker, I found a small launch button with its own speaker and battery that I could record on. I bought one for each of our grandchildren. For the sound I used Marvin the Martian from the *Bugs Bunny* TV show saying, "Where is the kaboom? There was supposed to be an Earth-shattering kaboom!" I gave Dan Osborne one that said "5, 4, 3, 2, 1, fire; OH, RUN FOR YOUR LIFE."

Eldon was proud of the second water well at Poker Flat, the one in the blockhouse. The other well was in Poker Inn, a building that housed a small kitchen, restroom with showers, and rooms for crew members to sleep in during the launch seasons. Eldon showed me the water softener system they had had to install because the blockhouse well water smelled bad. Eldon pointed out a chunk of beaver-gnawed wood that they had brought up from far underground as they were drilling the well for the blockhouse. Eldon and Neil told me the crazy story of how after an all-nighter launch attempt they had continued to work together trying to get well water working for Poker Inn. They had hammered a well point into the ground, knowing the water table was only a few feet below, yet had gotten nothing. They wondered if the well point had gotten clogged with silt. They attempted to clean it, and soon after, when they powered up the electric water pump, they were both sprayed with icy cold water. They were exhausted, but they broke out laughing together at their success. This story was only one of many I heard about the can-do attitude of Neil, Eldon, Larry, and the crew at Poker Flat and the skills they brought to bear to solve problems.

Neil told me that the blockhouse had come about because he had NSF funding for a telemetry system, and then Sandia had donated a nearly complete telemetry system. NSF had agreed the funding could be used to build a blockhouse instead. Although the blockhouse was a cast concrete building, it was listed as a "temporary equipment shelter" due to the Board of Regents policy that stated only the board could authorize a building. "Temporary equipment shelters" were exempt. So rather than wait for board approval, none was asked for. I saw the Sandia telemetry system in the blockhouse and the two auto-track 220 MHz antenna systems on concrete pads just outside the blockhouse that NASA Wallops had donated to Poker Flat. The blockhouse had a 2-foot-thick roof with another 4 feet of dirt piled on top of that. Though it was less than 300 feet from the launchpad, I was told that it should be able to take a direct hit of anything we were capable of launching and people inside would not be hurt.

Eldon and Neil then drove me southwest, away from the launchpads and blockhouse, to where NASA had installed one of their VERLORT radar systems. NASA had adapted a World War II S-band radar to track radar beacons for their first Mercury spacecraft efforts. The VERLORT was touted as being capable of tracking a radar-beacon-equipped payload to the moon. The VERLORT was in a specially built trailer. Poker Flat had built a small poured-concrete building about 8 feet square by 10 feet high that the VERLORT antenna sat on. Outside but near the VERLORT trailer stood an open-sight "naval gun director" platform. One of the NASA VERLORT folks would be stationed outside during a launch to optically track the rocket with this open sight as it lifted off. If the rocket failed, the open-sight operator would have time to step behind or into the concrete building that the VERLORT antenna was mounted on.

Across a large parking lot from the VERLORT radar stood the long flat-roofed Poker Inn. It was some 30 feet wide by a little more than 100 feet long and, because of permafrost, sat on pilings a few feet above the ground. It was constructed from four of the 12-by-50-foot-long modular buildings that had been temporary housing attached to the old Geophysical Institute, the Chapman Building today, on the University of Alaska Fairbanks campus.

The last thing Eldon and Neil did during my first tour of Poker Flat was to take me up the crazy hillside switchback road to where the army buildings for the Meteorological Rocket Network (MRN) tracking and data systems were under construction. I saw only two 22-foot radomes (short for radar domes, used to shield radars and sensitive equipment from the environment) and a 50-foot radome in place over concrete foundations near the top of the hill that day.

Larry warned me that Neil had not gotten new permits for impact of the rockets launched from Poker Flat from the BLM or the FAA for the coming year. When our administrative assistant Judy Holland gave me copies of both permits, I discovered that our paperwork for permission to launch rockets into air space had also lapsed. I went to the BLM offices, which were next to the Lacey Street theater in downtown Fairbanks. I held out last year's permitting document and asked if that person was still there. I expected bureaucracy, but I found this BLM office for the whole of northern Alaska had only a few really great, friendly people in it. I walked out less than a half-hour later with a signed permit to land rockets anytime in the next year. The previous permits with BLM were only for a block of land north of Fort Yukon where the upper stage and payloads would impact. My contact at BLM suggested we add land between Poker Flat and the Yukon River along the rough magnetic north azimuth, which he did.

By the time I retired from Poker Flat in 1989, we had 11 annual permits in operation: a permit with the FAA and permits involving landing rockets on multiuse lands overseen by Native corporations, BLM, State of Alaska Department of Natural Resources, and others.

Our FAA contact was Ed Keene, who worked out of a small office along the northeast side of the runway at Fort Wainwright. Ed was fine with our normal aurora borealis sounding rocket launches of a few per year into the polar overflight route that went directly over Poker Flat but told me the new request for permission to launch a metrological rocket every Monday, Wednesday, and Friday eastward for the US Army program was troubling. I was ill-prepared as he pressed his case forward, saying we should move our operations to the abandoned but still FAA-permitted Nike Hercules test range behind Eielson Air Force Base. Neil had told me that the area northeast of Eielson was way too small for future growth. I responded to Ed by stating our scientific reasons for the aurora rocket launches and said that they would occupy FAA-controlled airspace for only a few minutes—and that our launches would typically be late at night while over-the-pole commercial flights were always midday. I had absolutely no idea how to respond to his concerns about the meteorological rocket network launches. Truth be known, I had no idea what kind of rockets the MRN program had been or would be launching. I somehow managed to say that of course any aircraft takes priority over our launch attempts. If a plane is flying through, we will just wait until it is clear of the airspace that we want to use.

A couple of hours later, I was frankly amazed when Ed said he would sign off on a permit for the coming year and that his approval would guarantee approval from the higher-up FAA office in Anchorage. He added that so far there had been no incidents of near interactions between our rockets and airplanes, and that was good. Any future incident would be bad. I had no idea what had been done in the past at Poker Flat, but I told Ed that our roadblock people and other observers would also be outside listening and looking for aircraft. Almost as an afterthought, Ed added that of course there were always military aircraft missions that took priority, too, and that the FAA might call a hold for a launch from Poker Flat and not be able to tell us why. We should assume that it was because the military was using the airspace. Ed told me the FAA in Alaska worked with the air force and shared information such as the launch attempts from Poker Flat, which he called a "military due regard notice."

I had this same conversation with Ed every year for the next 5 years. He kept telling me he really wanted us to move Poker Flat. At about the same time that Ed retired, the FAA decided we needed to communicate with their Anchorage center, not Fairbanks. To my amazement, the Anchorage center never raised a fuss about us launching. I decided that because we had operated safely and not caused problems, and Ed had been in charge, they figured that even with him now gone, we were all right. We had an excellent safety record with FAA since day one at Poker Flat.

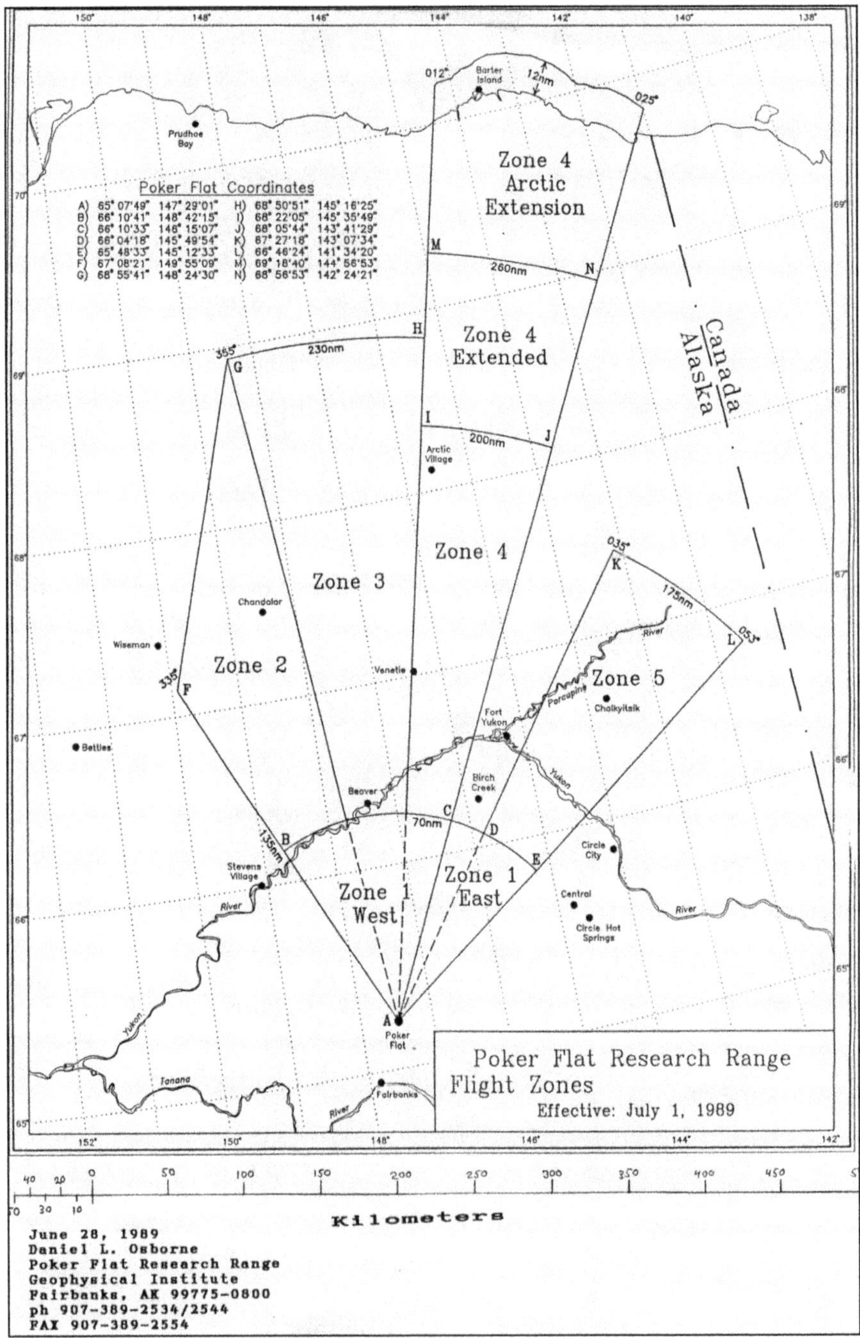

Each year the Federal Aviation Administration would provide to Poker Flat a map with permissible flight plans for pending rocket campaigns

Financing Poker Flat

The Geophysical Institute business office was a first-class operation led by Sue Johnson, who had assembled a great and very competent staff who were fun to work with. The business office helped principal investigators properly write the financial portions of their proposals, made sure the investigators bought only what they had proposed, and prepared and sent out monthly statements on how much money the investigators had left to spend before the end of their contracts. I learned the hard way that as good as the GI business system was, and as competently led as it was, the accounting system was inevitably a month late in its reporting. This epiphany occurred in the months following a work order I wrote to Norm Veach, the engineer who was hired to take the GI into the age of digital electronics. Norm managed to elude others and me as he spent our money and failed to deliver what we expected. I would up losing roughly $20,000 of Al Belon's air force contract money to build an upgraded control system for the Ebert scanning spectrometer system I was using to obtain near-infrared spectra from aurora.

I shut down the effort with Norm and complained to GI director Keith Mather and assistant director Neil Davis. When I started assembling the prelaunch science instrumentation for Ester Dome and Fort Yukon Observatories that would feed data directly to the blockhouse at Poker Flat, I quietly kept meticulous track of the expenditures for the project as I made them, both hardware and salaries. As a result, I found no surprises when I received the formal business office financial reckoning at the end of each month. I don't know if Neil or Sue were aware that I had developed this capability, but it did not take me much time to do it, and I had a much greater peace of mind about finances from then on.

Going into my first year as range supervisor, Neil told me that Poker Flat had a defined budget of about $400,000 for the fiscal year that started July 1 and ended June 30. Neil also told me that NSF, NASA, and Los Alamos Scientific Laboratory (today known as Los Alamos National Laboratory), all US government agencies, paid different amounts for Poker Flat to launch their rockets. NSF paid $5000 and the others $32,500 for each rocket we launched. The GI underwrote our operating costs but did not pay our salaries until after we successfully launched the rockets. This latter point worried me more than the former. It is a testament to these agencies' desire for Poker Flat to launch their rockets that they overlooked the federal regulations requiring government agencies to pay the same price for equal service, no matter what.

Before Neil left on his sabbatical in December 1971, he was asked by the Defense Nuclear Agency to submit a proposal for $1 million to upgrade the radar and other systems at Poker Flat. Neil called me into his office and showed me a ridiculously brief proposal he had written to DNA and had me sign as a co-principal investigator.

During my first tour of Poker Flat in August of 1971, I had been appalled to find that the rocket motors were stored in a military surplus Jamesway tent, a portable, insulated, semipermanent shelter originally developed for military use during World War II, particularly in extreme climates like the Arctic. The motors had to be kept about room temperature at all times and never be allowed to get cold, lest differential heating between the propellant, insulation, and metal casings cause a crack in the solid fuel propellant. If that happened, at ignition of the rocket motor, the flame could easily exploit the crack and cause a spectacular failure of the rocket. The interior of the Jamesway tent was kept warm with hot-water baseboard heating that hung from sheets of plywood along the length of the walls along either side. A small wood-frame building adjacent to the Jamesway held an electrically fired hot-water boiler. I was assured that, should commercial power fail, there were a couple of fuel-powered portable electrical generators on range that could be used to keep the motors in the Jamesway warm. In the year or two before my first visit to Poker Flat, Bill Baranauskas, one of the GI machinists, was living on range and would have kept the rockets warm if commercial power failed. But no one was living at Poker Flat now.

In consultation with the range crew, I decided we needed a more secure rocket storage building and one with an overhead crane inside to lift and move the rocket motors in their packing crates. (In the past, rocket motor crates had been pushed into or pulled out of the Jamesway with a small tractor. You don't even want to imagine how the crew had moved boxes of stacked rocket motor boxes across the relatively uneven dirt floor of that Jamesway.) Thinking the known potential income of $400,000 for the year would surely allow it, I initiated the construction of Rocket Assembly Building B (B-RAB), the new building we would use for several more years to store and build up the rockets.

Though I thought I was keeping good track of operating funds, that fiscal year I greatly embarrassed myself and everyone by spending a few thousand dollars more than we had building B-RAB. When it became clear I had created an overrun, Neil expressed his full support for me, pointing out that the philosophy of Poker Flat was to cost the university almost nothing. Together we picked up the telephone and called representatives of the agencies that had launched rockets at Poker Flat in March. I could not believe how readily each agreed to send us their fair share of money I had overspent building B-RAB. Neil just smiled as he called in some of those chips you talk about when you have done favors for others and now ask them to repay you in a reasonable way. During each telephone call, the representatives affirmed that it had been a great idea to improve the facilities at Poker Flat for safer handling and storage of their rocket motors.

My First Rocket Launch as Supervisor

My first rocket launch from Poker Flat after I was appointed range supervisor occurred at 0900 Coordinated Universal Time (UTC), August 27, 1971. Neil Davis and Larry Sweet took me to Poker Flat that afternoon so I could see final payload and rocket motor preparations and the launch. The principal investigator was Jim Barcus from the University of Denver. Jim was an intense, lean, well-tanned, chain-smoking man with an infectious grin and great sense of humor. He was a top scientist in his field and had flown many X-ray and high-energy-particle experiments onboard balloons and Arcas sounding rockets.

It was the 25th rocket launched in the history of Poker Flat: PF-BA-25, which stood for Poker Flat-Bullpup-Apache-25. Bullpup was the first, or booster, stage of the two-stage rocket system and Apache was the upper, or second, stage. Both boosters were "storable" liquid motors that used a hypergolic mixture of aniline (hydrozine) and red fuming nitric acid for fuels. They were surplused by the navy, which originally flew them on carrier-based aircraft.

The Bullpup was an unusual sealed liquid-fueled rocket. It was originally the first stage used to push a missile seeking targets off the wing of an airplane. The igniter breeched two chambers containing the hydrazine chemicals. Four huge fins were aligned along its short body.

This particular effort was one of the first rocket-borne experiments sponsored by the NSF. I first met Gil Moore of Thiokol, who became a lifelong friend, at this launch. Gil had arranged the use of the Bullpup-Apache system for NSF and Jim.

Communications with the FAA and everyone on range went well, and launch officer Carroll Coe pushed the firing button. The payload worked perfectly, and Jim and his graduate student, Roger Williamson, were very happy with the results.

I put Eldon Thompson and Lou Baim, with input from the rest of the range crew, to the task of detailing what we needed to design and buy as materials for Range Improvement Proposal #1. By this time we were putting Poker Flat's administrative assistant Judy Holland into major overload. I remember Sue Johnson spending a lot of time helping me craft financial data for this and other Poker Flat funding proposals. Sue was a really unique and charming person. She was extremely competent and always had a big smile on her face as she asked how she could help me. I fondly recall the occasional weekend day meeting with her at her home to work together before the financial data was ready to be sent to DNA. When Sue's husband wasn't home, she insisted that we move a table in front of a large window with the curtains wide open so the neighbors could see us.

One of the DNA-funded range improvement projects was to create a new entrance road into Poker Flat. We were going to place the new pad 1 launcher very close to the old entrance. We were working with Eldon's good friend Jim Thermin, who owned a local construction company, to get the road in place before freeze-up. Somewhere in the midst of this effort, Jim had done about $25,000 worth of work without a contract with us. Eldon assured me he knew all about change orders, and because of his friendship with Jim, it was not a problem. When Keith Mather got wind of this, he called an emergency meeting with Sue and me. I got read the riot act for allowing this situation to develop, and somehow we got hold of Jim and made everything legal. I was chastised and knew that I had let everyone down. After that, I was ever more watchful regarding finances. Nothing was ever said again about this screwup, although I'm sure Keith set Sue to watch me more closely.

Keith thought a lot of Sue and helped her get a fiscal internship at NSF. That would have been absolutely great for us when she returned, but she stayed at NSF in Washington, DC. Neta Stilkey took over when Sue left, and we successfully worked together throughout the rest of my time at Poker Flat.

During my tenure as director of Poker Flat, I felt then and now that Poker was a bit of cash cow for UAF. Our funding agencies were charged slightly less overhead than projects on campus, but still roughly 33% of the $1 million we brought in went to overhead. We got financial and administrative services from GI/UAF, but we had to buy our own toilet paper, clean our own floors, and do myriad other tasks that were typically done with overhead funding on campus.

My very first assistant I hired at Poker Flat was Randi Wagner. When Randi eventually moved on for another position in Fairbanks, I hired Peggy Dace Boyd, a friend of Randi's who had helped former UA president William R. Wood write public documents, as her replacement. When Keith asked me to write the 10-year history of Poker Flat, I panicked, but Peggy was unflappable. Between us we got the job done.

Years later when Peggy decided to take another job, I hired Pat Young to work in this position. The 25-year lease for the 2500 acres of land Poker Flat sat on needed renewing, and the UAF land office was under the gun with everyone for screwing up its stewardship and ownership of lands. Pat had extensive experience in land issues, first working for Ted Stevens, who was now one of our Alaska senators, and on land issues in the Arctic National Wildlife Refuge for Ave Thayer.

After Pat's tenure at Poker Flat, I hired Mary Farrell, who had extensive experience with spreadsheets, which we needed for our finances and other purposes. Mary worked for me until I retired from Poker Flat.

Determining Flight Requirements

During preparations for the DNA's Infrared Chemistry Experiment Coordinated Aurora Program (ICECAP) missions, Ed Allen of Space Data Corporation was hired to coordinate the project. In early January of 1972 I started hearing from Ed and from Los Alamos Scientific Laboratory, NASA, Air Force Cambridge Research Laboratory (AFCRL), and the Meteorological Rocket Network (MRN) team about what they needed to be able to launch their rockets from Poker Flat, especially in March. I was nearly paralyzed at times as I learned what they wanted and tried to figure out what we could do with the funding we had. I would not have survived this rapid onslaught of needs without the help of the crew at Poker Flat, as well as Judy Holland, Neil, and Larry in the moments he made available after his return to college studies.

Ed got his marching orders from the overall contract monitor at DNA in Washington, DC. He would fly to AFCRL in Boston, Massachusetts, and learn what was needed from Space Data Corporation in Phoenix, Arizona, and from Ed Butterfield, whose group at White Sands Missile Range supported the DNA's rocket work. Ed would coordinate and communicate these needs with us at Poker Flat, and then assess the progress we were making to be ready for the next ICECAP launches. Ed flew this circuit near constantly, called and talked, sent faxes, and produced and updated flight requirement documents for each ICECAP rocket experiment, because no two were identical.

Ed's efforts sometimes involved other contractors who supported the principals. Ed did this almost full-time during the first year of ICECAP, and for several weeks each year for each of the next few years of ICECAP launches.

The pace of communications and needs for the support of rocket launches for the 1972–1973 season also grew. By early November I did have some help in keeping track of what to do when Randi Wagner first came onboard as my assistant. Randi was a street-smart, well-trained person who could compartmentalize and not get confused by overthinking things, as I was inclined to do. Fortunately for me, Randi was also a great listener and clear thinker. She was the glue who held the various screwballs who worked there, including me, together on the same page, through focused communications.

Aside from DNA, none of the user agencies sent us a formal flight requirement document detailing all the needs, performance data for the rockets, the science objectives, and so on any earlier than a week before a launch. Despite the fact that these projects had been funded years in advance, last-minute changes always delayed the final product reaching us in time to ensure we were ready.

In the spring of 1973 I decided to emulate Ed Allen's methods, so I scheduled a trip to meet with the scientists and rocket folks to figure out what they needed us to accomplish at Poker Flat before they arrived. I took Randi with me. We spent a few weeks to a month that first spring and came back prepared to fine-tune the fiscal part of our annual proposal before our fiscal year ended on June 30. The trip was a huge success, and I repeated that annual flight-requirements-finding trip year after year until I left Poker Flat in 1989. Some quietly speculated that this was a boondoggle on my part, but when I reminded our funders that if they could get valid flight requirement documents to us as little as 3 months in advance, I would not need to travel, none volunteered make that happen. Quite literally, the government's bureaucracy and signature of approval lines of authority made it impossible for them to meet a 3-month deadline. In the end they all agreed that part of the reason Poker Flat was successful was because we worked so hard to meet the users' needs.

The Name Poker Flat Research Range

Soon after I became supervisor of Poker Flat, it occurred to us that we had no official name. As mentioned earlier, Poker Flat had been used internally by the people working on the project. Although Poker Creek is nearby, I was told the facility's name was actually a takeoff on Bret Harte's very short nine-page story "The Outcasts of Poker Flat."

Harte wrote about a mythical California gold rush community that banished a gambler and a prostitute from their evolving sophisticated town out into the winter cold. When the town's citizens reconsidered their actions and went out to bring them back, it was, alas, too late, for they had died. I was told that the crew who had met at the Geophysical Institute each morning in the fall of 1968 to drive to Poker Flat together considered themselves the institute's outcasts. In fact, they worked in pretty miserable conditions, accomplished a great deal, and earned the right to rally about this slogan. A not-quite-finished wooden plaque on the entrance to one of the main buildings at Poker Flat indicated that the typical practice of playing the card game poker while waiting for the proper auroral launch conditions might be another possible explanation for the name.

I accepted the name Poker Flat, but I worried about the implications of adding on the words "Rocket Range," as most others assumed it should be. When faced with adopting an official name, I persisted in my efforts to call it Poker Flat *Research* Range. I always considered it to be more than just a rocket launching facility. At the very least, it would be different than any other rocket range, because it was dedicated 100% to scientific research. I considered it to be a research facility and hoped we could someday achieve funding diversity as a result of our involvement in a number of programs. Through the years I worked hard to do that, with little actual success.

At first, NASA Wallops had listed our launch facility as FB, which stood for Fairbanks, in most of their activity publications. When we established our name as Poker Flat Research Range a couple of years later, they continued their FB designation. For several years when I asked that they change their listing to Poker Flat, I was told it was "difficult" for them. Many years later, activity reports from Wallops finally started listing our facility as Poker Flat Research Range.

Another problem involved the State of Alaska, whose map-making and road-sign policies kept them from allowing the name Poker Flat to appear on either. If I had a dollar for every Poker Flat-bound truck or car that has driven through Fox, Alaska, and on up the Elliott Highway, rather than turning onto the Steese Highway for Poker Flat, I would be a rich man. It seemed illogical that an internationally recognized scientific research facility, which often created spectacular displays visible throughout the Northern Hemisphere, couldn't be found on a single map.

Developing the Logo

I thought we needed a logo to represent Poker Flat, so Randi and I distributed a request for ideas to our own crew and to the GI staff and faculty. But when our self-imposed deadline arrived, we had only a couple of ideas, each of which used some variation of an Alaska moose in the foreground with our now-infamous A-frame launcher cover in the background. We wanted something more creative but were at a loss as to how to explain it.

A few weeks later Randi laid a logo on my desk. She had asked Al Anthon, a local graphic artist, for some ideas. He had given her only one, a pet idea that he had developed over the years. I liked its understandable but not quite map-like depiction of Alaska's position among the circumpolar nations. All we added was the name Poker Flat.

We bought a few hundred sticky-back logos with their bright red, white, and blue coloring. They were very popular with our range users. Our limited supply was quickly used up, and we had to buy more. They were so popular that someone suggested we put the logo on a T-shirt. We were cautious and bought only a few high-quality iron-on reproductions. Randi hung the first T-shirt up on the wall as an advertisement, and again we quickly sold our first order. One day I was looking at the T-shirt hanging on the wall and realized that Randi had carefully cut off the back. When I asked her why, she gave me her very logical explanation as someone who grew up New Jersey: people would steal it, if they could use it. It stayed on the wall and sold more T-shirts for years.

I bought Poker Flat T-shirts and gave them to my three children. Less than 2 years later, they wrote on a postcard from Paris that they had seen someone walking down the street wearing a Poker Flat T-shirt. Later, in 1976, when baseball hats with logos on them became popular, we got some of them, too, as well as sew-on patches and cloisonné pins.

Burning Rocket Propellant

Within a month of my appointment in August of 1971, Eldon Thompson brought a sliver of spongy rocket propellant into my office in the GI rocket lab. It was less than a quarter-inch in diameter. I could compress it between my thumb and index finger. You could see flecks of aluminum meshed into what appeared to be sponge rubber. I marveled at it, and he asked if I would like to see it burn. Since I had never seen rocket propellant to begin with, let alone see it burn, I said yes.

Eldon set it into a 3-inch-diameter film canister and set the canister in a heavy metal vise. He lit the propellant and sparks flew as it burned furiously. It burned longer than I imagined, and for some reason it fell off the vise onto the floor tiles. It continued to burn and I covered it with my shoe, hoping to contain it. I did, but there was now a half-inch-diameter hole through the floor tiles to the concrete underneath and a pretty good scorch mark in the bottom of my shoe.

Nike rockets are notorious for shedding a small amount of unburned propellant when launched. We often walked the Steese Highway in front of Poker Flat after a launch and picked up chunks a few inches long and an inch or so in diameter. We might find a dozen such pieces. If ignited in the open air, they burned smoothly almost without smoke and left behind only a tiny amount of ash. Carroll Coe told me the burn in open air was deceptive, and that in the closed container of the Nike steel shell pressure built up, which increased the speed with which the propellant burned.

In 1987 Carroll gave me several months' notice that he wanted to retire from Poker Flat. I had worried about what to do with four NASA Apache rocket motors that NASA said were not theirs, as well as a couple of hundred pieces of explosive ordinance used to cut bolts and initiate actions during rocket flights. I worried that vandals might break into our storage facilities to steal them, injure themselves, and sue us.

I had inventoried everything and submitted the list to NASA, the army, and the air force, but no one wanted the stuff back. Being hazardous, the stuff was hard to ship. So I asked Carroll to get rid of the Apache motors. He cut each into 6-inch sections using a water-cooled motorized hacksaw and then set each piece on fire. He built a fixed workstand and safely ignited all the small thrust motors and spin rockets left behind by previous rocket projects, as well as the ordnance.

Communications On and Off Range

Something like eight runs of spiral 4 communication cable were laid across on the ground between the top of Pedro Dome, where the US government had an extensive communications hub, and Poker Flat for the SECEDE III launches. Connections were made through this cable and Pedro Dome for the SECEDE III command center at Fort Wainwright to communicate with those at Poker Flat. Soon after the last launch that spring of 1969, these cables were picked back up, leaving people at Poker Flat with no way to communicate.

The Geophysical Institute had two UHF communication systems, each mounted in its own two-wheel trailer. I think these operated in the frequency range of 200 MHz, and each had its own communication tower with antennas. When I first visited Poker Flat as range supervisor in August of 1971, one of these systems was still in place near the blockhouse, though it was unused. At that time communications between Poker Flat and the Geophysical Institute were likely taking place through a 12 GHz experimental radio link that GI communications engineer Mel Holmgren had just finished installing that summer.

The communications station was a single, lightly populated electronic rack holding the half-watt microwave transmitter and receiver, connected to a 6.5-foot-diameter microwave dish. Stations were established at Poker Flat, Pedro Dome, Ester Dome, and in Mel's office at the Geophysical Institute. At Pedro Dome connections were made to the US government's AUTOVON lines and at the Geophysical Institute to commercial telephone lines. AUTOVON (short for automatic voice network) is a secure, priority-based telephone system developed by the United States Department of Defense (DOD) during the Cold War. It was designed to provide reliable, survivable communication for military and government operations in case of a nuclear event. This system functioned almost without fail through my 18 years managing Poker Flat.

For my first few years, we suffered seasonal outages of the microwave system, which we learned we could fix by tweaking the vertical alignment of the microwave dish at Pedro Dome aimed at Poker Flat. We convinced ourselves that this was occurring when the trees developed or shed leaves and attributed it to knife-edge refraction, as the microwave beam went just barely over a small ridgeline along the microwave path.

Someone came up with the idea of installing a passive repeater, which consists of two back-to-back microwave dishes, one aimed at Pedro Dome, one aimed at Poker Flat, with nothing more than a waveguide connecting the feed horn of one to the other. It was passive as there was no amplification involved, and most importantly in this case, it did not need electric power. In typical fashion, Eldon Thompson gathered the troops and hauled four tall electric power poles up a primitive road to the top of the ridge, where he drilled holes, set the poles, and had the crew build a platform on top of the poles large enough to hold the dishes firmly in place. He declared it ready almost before we finished talking about it.

Mind you, we did not even think about getting permission from whoever had physical responsibility for use of that land before we put up the platform tower. No one ever complained.

We had one but needed two 6.5-foot-diameter microwave dishes with feed horns for this project. We had another 6.5-foot microwave dish, and we took the feed horn from a smaller dish. We would have to find and mount the feed horn in a very exact focal position in the large dish, but initially none of us knew how we were going to do that. I was the one who figured out how to find the focal point. I stood the dish vertical in our Payload Assembly Building and, from some distance away, sat a small laser on a table, aimed along the axis of the dish. I took a small mirror and laid it on the surface of the dish in the path of the beam of laser light. The laser beam was relatively easy to see along its path and reflected from the dish because of dust in the air. We had to move the small mirror around on the surface of the microwave dish only a few times to determine the focal point of the dish.

We had no way to test if all this would work except by hauling the two dishes up and putting them on the intermediate tower, aiming one dish at Pedro Dome and the other at Poker Flat. Relatively quickly we saw an improvement in the signal strength at Pedro and Poker with just a few tweaks of the intermediate dishes. These were still functioning well when I left Poker Flat in 1989. We had by then made another major improvement at the Poker Flat end by moving our microwave terminal to the top of the hill behind Poker Flat.

A gentleman named Perry Stoop bought out commercial telephone service to the community around Pedro Dome a few years before I left Poker Flat. We had little faith he could provide all of the connection Poker Flat needed with the outside world by then, but we did buy one line from him, as did the Chatanika Lodge, about a mile and a half south of Poker Flat on the Steese Highway. Perry's line to Poker Flat worked fine, but he never got more than one line to us while I was in charge.

Computing Wind-Weighting

Soon after taking on the range supervisor responsibilities for Poker Flat, I felt that one of the most important things I needed to learn was how to take into account the wind that an unguided sounding rocket is going to fly through, in order to determine the launcher settings for it to fly and land where desired. In addition to the winds aloft, two other factors influence rocket flight: (1) the Coriolis effect, caused by the Earth's rotation that pushes north-moving objects to the east, and (2) the larger radius of the rocket's track above the Earth during flight, both of which tend to make a rocket land to the west because the Earth is turning to the east under the rocket.

Everyone, kid or adult, who builds and launches model rockets quickly learns how to compensate for wind drift before they launch so the parachuted model rocket lands where it is easy to recover and not, for example, in a tree. Wind-weighting a model rocket is sometimes called "weather cocking." Contests are held with the objective being to land the parachuted model rocket as close as possible to a particular point on the ground.

Typical model rocket motors burn for only a fraction of a second, and thereafter the model rocket will drift in the direction the wind is blowing. After the parachute is deployed at apogee, the parachuted payload will drift in the direction the wind is blowing as it comes down. Sounding rockets used for studies of the aurora or Earth's atmosphere are just bigger, heavier, and burn longer than model rockets.

But the biggest difference in response to wind between model rockets and sounding rockets is caused by the longer burn time (acceleration) of the sounding rocket. A thrusting sounding rocket motor has an equal and opposite upward reaction responding to the downward force of the burning propellant, and if there is a side force from the wind, instead of drifting with it as a burned-out model rocket does, the sounding rocket will dive into the wind. To find the angle of flight of the rocket, it is necessary to mathematically resolve these two forces. And this flight angle is a constantly changing number as the thrust of the rocket varies and the winds change in speed and direction as a function of altitude.

The rockets launched at Poker Flat typically travel their first 24 feet along a steel guidance launch rail. They accelerate quickly and clear the tip of the launcher within one-tenth of a second at speeds in excess of a couple hundred miles an hour. Even though initially guided by the launch rail, these rockets are accelerating and the force of surface winds typically cause the most change in the aiming point within a few hundred feet of the ground. Of course the rocket is quickly accelerating and experiencing higher forces, so that the higher in altitude it goes, the less effect the force of the winds will have.

For this reason, it is important to measure surface winds accurately and take their affects into account by mathematical calculations right up to launch ignition. There were six levels of wind speed and direction measurements made on the 265-foot-tall meteorological tower next to the blockhouse and launchpads at Poker Flat. Data from these alone typically provided about 90% of the information needed for the rocket to fly toward a given aiming point, referred to as three sigma.

At Poker Flat, three concentric circles of probability were calculated for each flight. There were three concentric circles, with (1) sigma being the center and smallest and where 85% of the rockets would impact; (2) sigma was slightly more than twice as big as one sigma and is where 94% of the rockets would impact; and (3) sigma is more than three times as big as one sigma and is where 97% of the rockets would impact.

Since downrange safety was of major importance, winds above the 265-foot tall tower were measured often. A pibal, a 3-foot-diameter helium-filled balloon with a tinfoil bowtie radar reflector tied to it, was launched and tracked to determine wind speeds and directions from the surface to 10,000 feet every 20 min. A larger 10-foot-diameter helium-filled balloon with a more sophisticated radar corner reflector was launched at least once each 4 h to determine winds from the surface to 100,000-foot altitudes. The upper altitude winds did not change as quickly.

By and large, wind-weighting sounding rockets work well. Gusting winds will cut down a mission, as they are too variable. A no-wind aiming point is the point at which the rocket would land if all the winds suddenly and inexplicably died. The no-wind aiming point can never be where people live. Azimuth-angle wind corrections are more easily accepted than elevation-angle wind corrections. If the winds are too strong from the launch azimuth, the payload will pitch down in elevation angle too much to reach the desired altitude. On the other hand, a tailwind might cause the flight elevation angle to pitch back overhead south toward Fairbanks—also a no-no.

The rocket motor performance is usually pretty well-known and documented. But the thrust of the propellant may have some slightly off-center components. Another variable can be thrust misalignment created by misalignment of the hardware that holds the rocket parts together. Wind shears at ignition altitude for an upper stage can also affect alignment and performance.

Whereas model rockets can reach altitudes of a few hundred to a thousand feet, sounding rockets typically reach 62-mile altitudes at a minimum and can reach maximum altitudes of several hundred miles. Before launching sounding rockets, a great effort is made to find a safe landing point on the ground a few hundred miles downrange—safe as in not having the predicted impact point be close to where people live.

I met with Poker Flat's wind-weighter Jim Woolf and asked him to explain the process to me. Like Carroll Coe, Jim had retired from the US military to take this job at Poker. Jim had done rocket-related meteorological work in the army and had been part of the group that had done the wind-weighting for the first rocket launches from Poker Flat.

Jim had a large acetate-covered chart on a drafting table beside his desk, and he showed me how he marked it with wind and rocket performance data to calculate launcher settings. Local wind data for the SECEDE III launches had been acquired by filling and launching small weather balloons with lit road flares hanging beneath them. Two observers and helpers a few hundred yards apart on range measured and recorded azimuth and elevation angle and time data with tripod-mounted surveyor theodolites. He said data on winds from the surface to 100,000 feet altitude for SECEDE II had been derived from the Weather Bureau balloons that were launched at noon and midnight Coordinated Universal Time from National Weather Service facilities at the Fairbanks airport, although I also got the idea that perhaps they had temporarily installed a balloon system tracker at Poker Flat for SECEDE III.

Jim enthusiastically drove me to see the balloon inflation shelter that had recently been relocated to Poker Flat. It was a good-sized small building with a tall overhead door. Balloons could be inflated inside and then taken outside for launch. I recognized the building right away: it had stood on campus and was part of preparations the Geophysical Institute had made to support balloon-borne research investigations of aurora.

Balloons were now launched from this building at Poker Flat and tracked by the crew of the NASA VERLORT radar system. They printed out the track record, and Jim would pick it up and bring it back to the blockhouse. He interpreted the radar track data for wind speeds and directions as a function of altitude and used data from the two wind speed and wind direction units on the 65-foot-tall meteorological tower near the blockhouse and entered these data graphically on the acetate that lay over a standard computing form on his drafting table. He showed me how he got and entered rocket performance as a function of altitude data, ultimately graphically deriving the launcher settings. There was obviously some math behind all of this, but Jim's calculations were all accomplished graphically.

Jim told me that for the purposes of calculating launcher settings for a particular rocket, you would ideally have an anemometer to get wind speed and wind direction data on each launchpad and get anemometer data from the meteorological tower and data from a balloon launched and tracked to 1.8 miles each 20 min and to 18.5 miles each 4 h. He pitched to me the need for a taller and better-equipped meteorological tower. We did get a better-instrumented 265-foot-tall meteorological tower in place near the blockhouse and working before the first major launches of the following winter in March of 1972.

I asked Jim just how long it took him to do a launcher setting calculation after getting a wind field update corresponding to the flight of the balloon to 1.8 miles. He took me through the possible amounts of time to update the calculations, which were of course dependent on how much the wind field at the surface to 1.8 miles had changed. I suggested that this sounded like about 5 min of work if typical wind field changes had occurred, and Jim concurred.

Neil Davis bought and used an early minicomputer, an Interdata Model 4 with 4 kilobytes of memory, to do television track work during SECEDE II at Eglin Air Force Base in Florida. When he no longer needed it, he gave it to us to use as a wind-weighting computer at Poker Flat. By then we had grown to have four major launchpads available. Each launcher might have a different set of rocket motors configured to carry its payload into the aurora, and each set of rocket motors on a given launcher had its own unique set of performance data that had to be folded in with wind-field data to get launcher settings. While Poker Flat could support up to four completely different science experiments, one on each of its launchers, it was basically one launch at a time because we had only one radar-tracking system and one telemetry-recording system.

Jim Woolf announced that he did not want to learn how to use a computer to do his job. I tried to convince him otherwise. He was such a nice guy and so easy to get along with in so many ways that the strength with which he said this surprised me. To me, using the Interdata Model 4 was but a baby technological step in Poker Flat's future capability to support the safe launch of sounding rockets. I forecast that with a computer we could get data from the NASA radar electronically and more quickly, but Jim was not willing to change, and so I had to let him go. We lost a really fine person, and many of the first crew later voiced their concern that I had made a mistake.

The first electronic connection we accomplished after installing the Interdata Model 4 and programming it to calculate launcher settings was to the anemometer units on the meteorological tower. Merritt Helfferich was the first to wind-weight a rocket launched from Poker Flat using the Interdata Model 4. When Merritt moved on, we hired Tim Eggley, an ace electronic technician who worked with us for less than a year. Dan Osborne did wind weighting for about 6 months until we hired the next wind-weighter, Henry Cole. Henry had completed everything for his physics doctorate from the Geophysical Institute except typing up his thesis. He needed the job and did it well. Henry often got the computer keyboard greasy as he ate fried chicken, so we bought a cover for the keyboard. Henry often sang wonderful songs with DNA's rep Jim in the blockhouse. We all enjoyed Henry. He wanted to stay and do this job forever, but I strongly suggested he not. He went on to become a trusted science advisor to Alaska's governor Stephen Cowper and has held many much better positions since leaving Poker Flat.

At some point we obtained a small World War II gun-laying radar from White Sands Missile Range. Bob Spies installed and configured this small radar to track the balloons we launched every 20 min and connected it directly to input data to the Interdata Model 4. Eventually we moved on to a newer, more capable Nova minicomputer system. We were using Novas on our meridian-scanning photometer systems on our scientific data systems at Poker Flat and Fort Yukon.

We needed to hire a new wind-weighter, and I wanted this person to be able to get us into the future of using microcomputers for other applications on range. At the time Charley Lasater was our electronic communication technician. Charley had appeared for his interview in colored sunglasses and a polyester suit. His appearance really put me off, but he had great recommendations from his work at the European Space Research Organization (ESRO) tracking facility along Chena Hot Springs Road. The ESRO facility had closed down, and Charley was looking for a job. He had extensive experience with telemetry equipment, and we were hoping to take over the DNA-funded Space Data Corporation telemetry systems at Poker Flat. We prepared for this possibility by hiring Charley. In a surprising move, DNA not only gave all of the telemetry gear at Poker Flat to NASA but helped NASA upgrade it and helped fund the future operations of this same equipment. By this time Charley had more than adequately proven to us he was a real down-to-earth guy who would do almost anything and do it really well. He was a very knowledgeable electronic technician.

I hadn't found anyone already working at the GI who wanted to play with microcomputers. I happened to mention my interest to Charley, and he said there was a really good electronic technician at ESRO who was always playing and building microcomputers. I interviewed Mike Cogan and was very impressed. Mike was born and raised on a homestead on Haystack Mountain, just west of Poker Flat. He was in his late 20s, had many brothers and sisters, and was helping his family take care of one brother with mental problems. Mike was single and clearly very religious. He did not smoke or drink. He was hard-working. He knew electronics and microcomputers extremely well, even though he had no schooling beyond high school. He had excellent recommendations from ESRO and the Fairbanks Memorial Hospital where he had worked.

Poker Popcorn

There were two kinds of "Poker popcorn." During the first few tense years of launch operations, we called aspirin Poker popcorn and always kept a full bottle on launch controller Randi Wagner's desk. Everyone knew she had it and would ask her for some when they had a bad headache from working the dozen or so nights before we finally launched a rocket.

The other popcorn was real, and we ate a lot of it. It was made in the blockhouse. When we were not counting below T-minus 5 min, which was most of the time, crews from telemetry and radar sent a representative over to pick up a gallon of freshly popped corn for their team.

We cooked popcorn in oil in pans. It filled the air with the smell of freshly popped popcorn, along with mists of oil and butter. The memory cores of our Interdata Model 4 computer were out in the open on the back. One day Joel Lindsey, one of our electronic communications technicians, decided to clean the popcorn grease from the cores. Each core was roughly half an inch in diameter and had two wire wraps at right angles to one another running through it. There were a couple of dozen of them. Joel used a very fine paintbrush, but to his horror he inadvertently broke many of these wires, which were vital to the computer's operation. Bob Spies initially went slightly livid with rage, then a tense hour or two followed as he soldered the teeny wires back to restore the computer's memory. I vaguely remember we missed being ready for the opening of the launch window that night, but the aurora gods held off from offering us an aurora fit to launch a rocket until Bob got it fixed.

In the next generation of popcorn popping, we wore out two or three hot-air poppers each spring launch window. A couple of generations later, our communications technician Ed Heath bought a commercial-sized popcorn popper, the type that is about 2 feet square and three feet tall with a heated tip-able bucket. It lives in the administration building, where there is less chance of damaging delicate electronics.

I hired Mike Cogan, and he excelled in everything he undertook and brought Poker Flat into the future electronically and with wind-weighting in many innovative ways. He could program in many computer languages and could also tell other software programmers what he wanted done.

After a few years, I got a call from the IRS: Representatives of the IRS and FBI were coming to Poker Flat within the hour to talk with Mike about his not paying income taxes. They asked me to not tell Mike they were coming. I told them that not only would I tell Mike but that I would insist on being present while they talked to him. They did not like my response. Mike was a very gentle person, always with a smile on his face, and wanted to help anyone do anything. The IRS and FBI came and in very measured but genial conversation made their point that Mike had to pay taxes, and while I don't know the details, I do know that Mike did. The feds were wearing guns, prepared to deal with a potentially hostile antigovernment person. Mike never raised his voice and they did not either.

One day I asked Mike if he could create a display for the flight data NASA radar was acquiring of rockets in real time. It did not take Mike long to do so because NASA recorded their data on magnetic tapes and kept copies of those tapes. A few days later Mike called me in to see it work. There on the screen we watched two plots take place. One was the flight azimuth versus time plot, and the other showed the altitude versus time plot. It was especially fun for me to see the flight azimuth versus time plot in real time.

For two summers in a row, I sent Mike to NASA Wallops Island to learn more about wind-weighting. After the second summer, NASA Wallops sent a woman named Jenny, who was about Mike's age, to Fairbanks. Jenny and Mike had met at Wallops, and a spark of interest had started. They eventually married and lived on the Haystack homestead.

The Road to the Top

When I took over at Poker Flat in 1971, the MRN program was in the process of moving from Fort Greely to Poker Flat. The MRN was created to coordinate synoptic weather conditions using sounding rockets. In this context, synoptic referred to synoptic meteorology, or the large-scale study of atmospheric conditions across broad geographic areas.

The Poker Flat crew was building a road from the electrical substation area to the hilltop where the MRN program would be located. That summer the US Army Atmospheric Sciences Laboratory at White Sands Missile Range contracted with the Geophysical Institute to prepare the hilltop site and to take down, transport, and re-erect a 55-foot-diameter radome in which they would install systems for the control of the Nike Herc and Loki-Dart rocket programs' telemetry tracking systems.

I was told that the steep switchbacks on the upper part of this road were not planned. It seems that one cold day Eldon Thompson and Larry Sweet had been blazing a path up the hill for this road on foot, with a bulldozer clearing brush following behind them. While trying a couple of different paths through the brush near the top of the hill, they suddenly realized that the bulldozer was right behind them, with the operator doing the best he could to make sense of the erratic path they were leaving. That road continued to be used, although we did build a straight steep addition to the west of the switchbacks for those instances when we had to bring 40-foot trailer vans to the top of the hill.

When I visited the hilltop site with Larry in August 1971, I saw a broad expanse of cleared land. The contractor who had been hired to do site preparation had scraped a foot-thick veneer of subarctic tundra and trees off of the tough underlying Birch Creek schist rock. The rock drilling tools and boxes of dynamite still scattered about were mute testimony to the difficulties the contractor had had in preparing the site thus far.

Two 22-foot-diameter concrete cylinders, each about 8-feet high, sat on concrete pads to either side of one 55-foot-diameter concrete pad with a foot-high cylinder of concrete around its outer edge. I saw various internal pillars and precision sidewall holes in all three of these cylinders. Two other 20-by-30-foot pads of concrete had been poured in the area.

Ralph New told me that the bolt circle that would be used to tie down a 55-foot radome was not right and that Larry had hired GI machine shop leader Jim Parry to design, fabricate, and install adapters that turned the contractor's misfit system into the perfect circle of bolts needed for the dome.

While it was initially thought that it would be easy to dismantle a 55-foot fiberglass radar dome from the deactivated Moose Creek Nike Herc battery 25 miles east of Fairbanks near Eielson Air Force Base, move it, and erect it at the MRN hilltop site at Poker Flat, it turned out to be not so simple. The GI crews ran into problems. The rope-like sealant between the radar dome panels tore erratically as they were taken apart. An intermediate step of laboriously peeling the sealant off the edge of each panel was necessary before they could be reassembled. A new plan involving two crews was devised: one crew would drive 25 miles east each day, pull panels at Moose Creek, and bring them back to the Geophysical Institute, where they would peel off the sealant and reload them on their truck. The next day another crew would drive that truck 30 miles north to Poker Flat loaded with cleaned panels to the assembly site. This actually worked quite well. Considering that none of the GI people involved had ever actually seen a dome being built before, the work they did taking it down, moving, and reassembling it was pretty amazing.

Ralph came up with the idea of adding a pulley outside at the very top of the 55-foot radome with a loop of lightweight rope running through it that reached the ground. Ralph said he would use it to knock off any snow that might pile up on the roof.

Larry Sweet told me the army had not come up with enough money to drill a well, so the contractor was going to install tanks to store water and a septic system.

A civilian named Alton Duff was in charge of the army meteorological program. Alton worked with Ralph and the Poker Flat crew to install the insulated panels into a raised-floor circular building inside the 55-foot radome. They then installed the Nike Herc target-tracking pedestal in one of the 22-foot concrete cylinders and installed a radome over it and a Nike Herc missile-tracking pedestal in the other 22-foot pedestal and installed a radome over it. In the building inside of one of the radomes, they installed the Nike Herc control and operating electronics and the Loki-Dart telemetry receiving control electronics. Sometimes Alton overstepped, trying to do things that were not in our contract. He pushed so hard a couple of times that I said, in no uncertain terms, "Hell no." He blustered and threatened, but nothing ever came back to me from higher up. We continued to do what we were formally committed to do for the improved MRN program, and more.

We laid a concrete pad about 300 feet east of the blockhouse and helped Ralph and his MRN crew install a boosted Arcas and Loki-Dart launcher there. The army also moved a 10-by-30-foot metal-clad building near these launchers to store rocket motors and payloads. The Dart payloads were also prepped in this building before launch. The boosted Arcas and Loki-Dart rockets were ignited by army systems installed in the Poker Flat blockhouse.

The Arcas rocket was one of the early solid-fuel rockets built for general purpose investigations of the Earth's upper atmosphere, up to about 50 miles above the surface of the Earth. The rocket payload experienced no more than about four Gs of acceleration, which meant you could build a payload in your basement or garage that would easily survive rocket ignition and the ride to altitude. Once at altitude, a small explosive train near the head cap of the Arcas fired and pushed off a standardized parachute canister and battery-pack radio transmitter telemetry package that had been beneath the nose cone.

The parachute canisters for three Arcas rockets that Willis Web would launch at Fort Greeley for his noctilucent cloud program had water in place of parachutes in the parachute canisters. Two of the three water-carrying Arcas rockets were launched. Neither created a noctilucent cloud. They were equipped with a simple light sensor. When deployed at altitude, the light sensor would confirm whether or not the rocket had gotten into direct sunlight, a condition that exists for natural noctilucent clouds. Noctilucent clouds are typically in a very narrow altitude range near 51 miles. They are easily visible only when illuminated by the direct glint of light from the sun when the sun is between 6 and 16 degrees below the horizon.

The easy ride for the payloads of the Arcas rocket system meant the Arcas was very susceptible to being offset from its desired flight path by local winds as they accelerated. An adaptation was the boosted Arcas system. An explosive charge was placed in a chamber with a booster fit against the rear end of the Arcas. This whole assembly was enclosed and elevated to launch angle. The explosive charge was fired first to push the Arcas up and out. Before leaving the launch tube, the Arcas was also ignited. This got the system moving faster and made it somewhat less affected by local winds.

The small production of the wind-sensitive, relatively expensive Arcas rockets gave way to the inexpensive (when produced in large quantities) Loki-Dart. But more important was that the Loki rocket burned its fuel in less than 2 s to achieve an altitude of about a mile, and then momentum took it to an altitude of 50 miles in a couple of minutes. This resulted in some 160 Gs of acceleration. Local winds did not affect the performance of the Loki rocket motor. The Dart looked like a rocket but had no propulsion. A slow fuse, called a powder train, in the Dart was fired just before liftoff of the Loki. Two minutes later, the powder train fired the ejection charge, which pushed a radar-reflective parachute, a temperature sensor, and a radio transmitter forward inside the Loki's tubular body with enough force to dislodge the solid steel nose cone, thus deploying the science package. The Loki flew very fast, and nose cones found later were often ablated from high-temperature heating. A standard Weather Bureau ground meteorological direction finder (GMD) tracking system received and recorded the temperature sensor data and the azimuth and elevation angle of the science package.

I was always very big on public relations, especially with the public schools in Fairbanks in the 1970s. The Fairbanks North Star Borough School District had funding for buses to take students to interesting places. I'm not exaggerating when I say that hundreds, perhaps thousands, of elementary students rode their buses to Poker Flat and observed Loki-Dart launches between 1972 and 1978. Ralph and his replacement, Bill Badger, held more than one launch for a few minutes while we offloaded and got 40 or so students into position to observe the launch from the far side of the dome. They heard the countdown and then saw the streak of smoke left by the Loki. We did a little show-and-tell with each group about what was being studied, and they were always very impressed.

Army MRN Program Benefits Other Science

Through these years there were some fun science programs that used the Poker Flat MRN capabilities. Dick Goldberg at NASA Goddard Space Flight Center developed an extensive two-year-long, multirocket, multiscientist program to see if the ozone layers above Poker Flat were changed by high-energy aurora events that typically occurred in the early morning hours. We launched rockets when our riometers indicated significantly enhanced D-region absorption of galactic radio noise. Some of Dick's rocket-borne experiments flew aboard standard Nike-Tomahawk sounding rockets, some aboard boosted Arcas, and several aboard Super Lokis with specially instrumented 2-inch-diameter Dart payloads.

In late 1977, the Soviet Union notified the United States and Canada that the nuclear power pack for one of its satellites was potentially going to land in northwestern Canada. These power packs were typically separated and boosted with a rocket motor to altitudes where they would remain for thousands of years. Something had gone wrong. Indeed, NORAD Space Command watched the package reenter the atmosphere, and search teams were deployed to find it. They were unsuccessful. It turned out that one of the routine Poker Flat MRN data sets, acquired by analyzing the rocket-deployed parachute at a time close to the reentry, detailed a stratospheric warming event. It measured stratospheric temperatures of around 0° Fahrenheit rather than the typical temperatures of minus-50° Fahrenheit. Zero-degree air is less dense than minus-50-degree air. When the people searching for the Soviet nuclear power pack recalculated its path through this stratospheric warming event, which covered several degrees of latitude and longitude, they quickly found it. Once they got close, its intense radioactivity led them right to it.

We worked with Ben Balsley with the National Oceanic and Atmospheric Association in Boulder, Colorado, to establish a mesospheric, stratospheric, tropospheric radar system in a gravel pit just up the Steese Highway from Poker Flat. It was an experimental system that could be used to acquire wind speed and direction data 24/7. The MRN team did a number of measurements simultaneously with operations using that radar system, and the in situ measurements from Poker Flat confirmed the accuracy of mesospheric, stratospheric, and tropospheric radars. More were built and deployed in other parts of the world.

My great on-range role models for how to grow a crew of competent people who worked well together were Carroll Coe, Ralph New, and Bill Badger. Ralph, like most of the people at Poker Flat, exemplified a can-do attitude. He knew how to get things done, officially or not, and he took great care of the enlisted army folks he worked with. Ralph retired from the army in 1975 and was asked whom he thought might do a good job replacing him as leader of the MRN at Poker Flat. He helped his good friend Bill Badger, who was stationed in Seattle at the time, get the job. Bill, like Ralph, held the rank of chief warrant officer 4 in the army. Bill was also a super nice hard-working guy.

Both Ralph and Bill created a real esprit de corps for their troops. The enlisted men of the MRN team brought their wives and families to Poker Flat for picnics each summer and invited the Poker Flat crew to join them. We all shared a willingness to do anything to make Poker Flat work. I knew when I took the job as supervisor that this attitude was something to be cherished and transformed so that we did not continue to offend the management or our co-workers in the Geophysical Institute. Having a budget of almost nothing and borrowing everything and returning it broken would become a thing of the past during my tenure. While Poker Flat continued to borrow equipment from the Geophysical Institute as needed, we made a conscious effort to make sure it went back in as good condition as it was when we borrowed it. On several occasions when we broke or inadvertently misused items, we bought new items as replacements.

Photos

Neal Brown in 1963 in the William Elmhirst Duckering Building at the University of Alaska in Fairbanks where, as a graduate student, he taught an undergraduate introductory physics class

Poker Flat Research Range Supervisor (later Director) Neal Brown in 1972 with Poker Flat crew: (left to right) Dianne Lewis, Eldon Thompson, Joel Lindsey, Beau Battey, Jerry Duncan, and Carrol Coe

Neal Brown at NASA Ames in 1961

Kristopher Brown, Nathaniel Brown, and Melody Brown Burkins with the first-stage nozzle of a sounding rocket that fell back onto Poker Flat property in February 1977

Neal Brown in 1975 outside the Geophysical Institute at the University of Alaska in Fairbanks with his three children: (left to right) Melody Brown Burkins, Kristopher Brown, and Nathaniel Brown, ages 7, 8, and 5, respectively, when the photo was taken

Poker Flat Director Neal Brown and Eldon Thompson in 1977

Neal Brown inspecting a sounding rocket being placed on pad 3 in advance of a launch in February 1976

A wideband satellite receiving system, used to test effects of the aurora on satellite communications, was installed at Poker Flat in 1977

Neal Brown in front of a sounding rocket with a nose cone loaded with barium thermite, used to create an artificial aurora when released into the Earth's ionosphere during launch

An engineer at Poker Flat works at a Data General Nova workstation performing wind-weight calculations

Neal Brown with a Terrier-Malemute sounding rocket being readied for launch in the winter of 1978. The Terrier-Malemute is a two-stage sounding rocket that uses a solid propellant to launch payloads of up to 400 pounds to an altitude of approximately 220 miles

Crew of Poker Flat users posing in front of a sounding rocket being readied for launch in the winter of 1973

Research range supervisor Neal Brown, 45, has worked at Poker Flat for the past 12 years. Brown is standing in the Optics Site, which houses such instruments as tracking dishes in the background to gather in data from the rockets launched into the ionosphere.

Fairbanks Daily News-Miner Heartland *magazine article about Neal Brown and his work at Poker Flat*

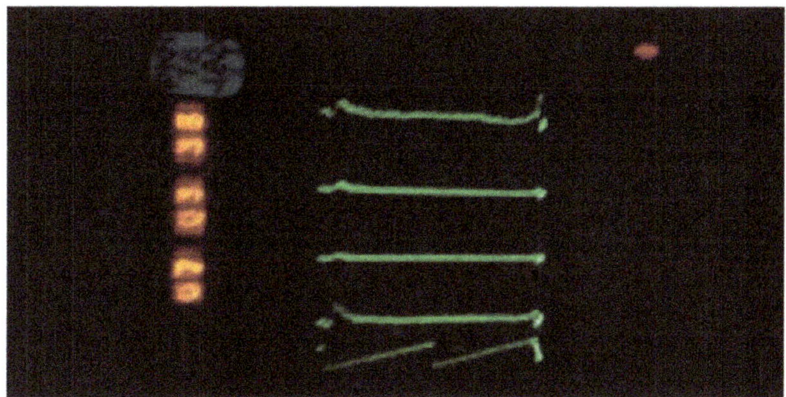

Beginning in 1976, Poker Flat scientists began using meridian scanning photometers, or MSP, instruments to measure the intensity of the various important wavelengths, intensity, and position of light from the aurora borealis to make go/no-go launch decisions

Bestselling author James A. Michener visited Poker Flat in 1984 and Neal Brown presented him with a Poker Flat hat

An engineer from the Geophysical Institute of the University of Alaska Fairbanks examines a sounding rocket prior to launch during the winter 1976 launch season

A dual-boomed platform that could be used for the construction of launch pads and systems was one of the first pieces of equipment that Poker Flat purchased in 1968

Crew members from Poker Flat review a fully loaded sounding rocket prior to a 1979 winter launch

Engineers from the Geophysical Institute mating the payload section to the second stage of a sounding rocket on pad 4 during the winter of 1979 launch season

To protect temperature-sensitive telemetry equipment in the payload of sounding rockets from weather fluctuations, sacrificial insulation is put in place and the rocket is launched, destroying the insulation. The work shelter on wheels is removed once full insulation is installed

Heated air is pumped through a series of tubes to maintain the temperature-sensitive solid rocket motors and telemetry inside the payload of a sounding rocket prior to launch at Poker Flat in the winter of 1979, when temperatures were routinely below minus 40 degrees

Monitors used to assess and record data transmitted from rockets launched into the magnetosphere to better understand interactions between charged particles from the Sun's plasma and the Earth's magnetic field

At that time maybe the world's most sensitive color camera, an Ikegami HL51S camera is used to record the aurora in real speed and colors in conjunction with a sounding rocket launch at Poker Flat, May 1985

An all-sky camera used at Poker Flat in 1973 to capture images of the entire sky at regular intervals, providing important insight into the local aurora

Sounding rocket payload being moved from the Payload Assembly Building to a launch pad at Poker Flat in the spring of 1984

Engineers and technicians in the Payload Assembly Building at Poker Flat during the winter 1980 launch campaign. This is where rocket parts were unpacked upon delivery and assembled for launch

Crew members at Poker Flat place a sounding rocket on a launch pad before a February 1980 launch to gather data about plasma interactions in the Earth's magnetosphere, where charged particles from the Sun interact with the Earth's magnetic field and atmosphere

An engineer from the Geophysical Institute performs a final inspection of a fully loaded sounding rocket just before launch in February 1980

Argo D-4 Javelin sounding rocket being launched from Poker Flat in 1982 to study the Earth's magnetosphere. Photo by Neal Brown

A Nike-Orion sounding rocket being launched from Poker Flat in 1979 to study the Earth's magnetosphere. The Nike-Orion is a two-stage solid-fueled rocket used solely for scientific research. It combined a Nike booster (a surplus from the U.S. Army) with an Orion motor, allowing it to carry scientific payloads of up to 150 pounds to an altitude of 118 miles

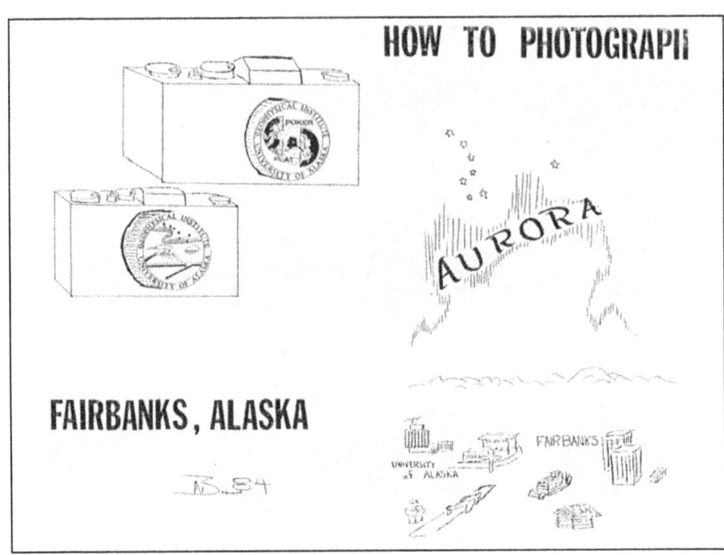

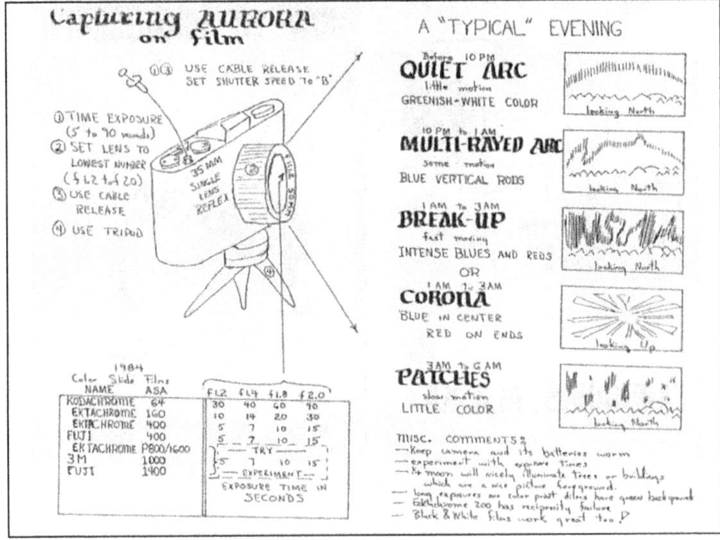

An avid photographer, Neal Brown would often teach courses to tourists and others on how best to photograph the aurora. He developed and created these posters in 1984 to help explain his methodologies

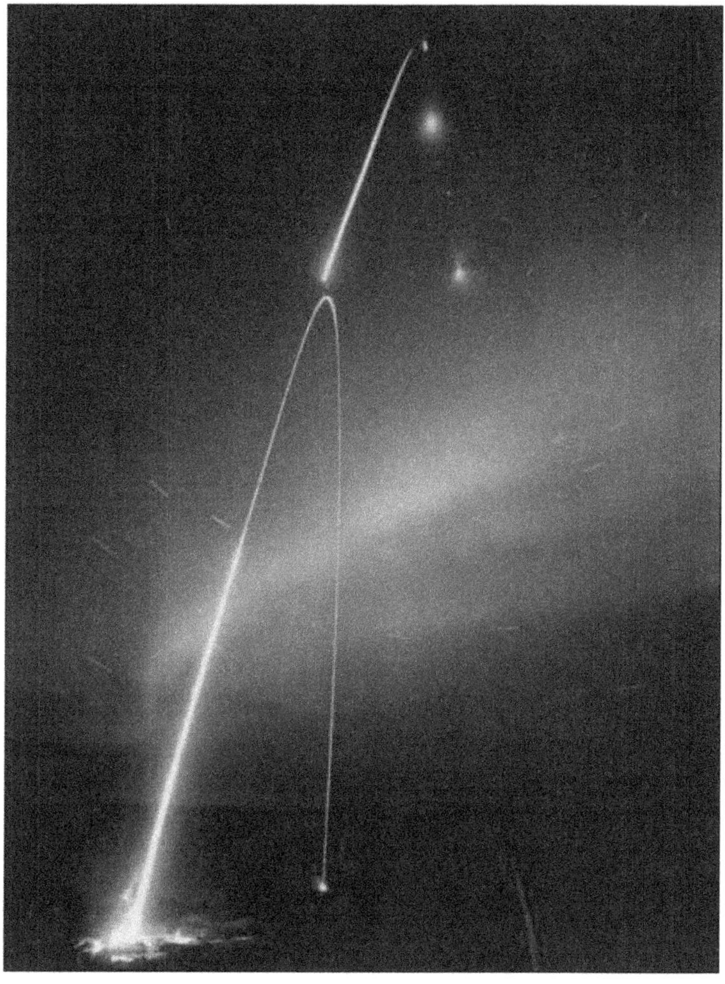

Long-term exposure photograph by Neal Brown, showing the launch of a two-stage sounding rocket from Poker Flat in February 1981, including the return of the first-stage booster

An engineer from the AFGL and USU tests equipment to be housed in the payload of a sounding rocket to be launched in February 1980 to assess the correlation between upper atmospheric magnetic activity and the aurora borealis

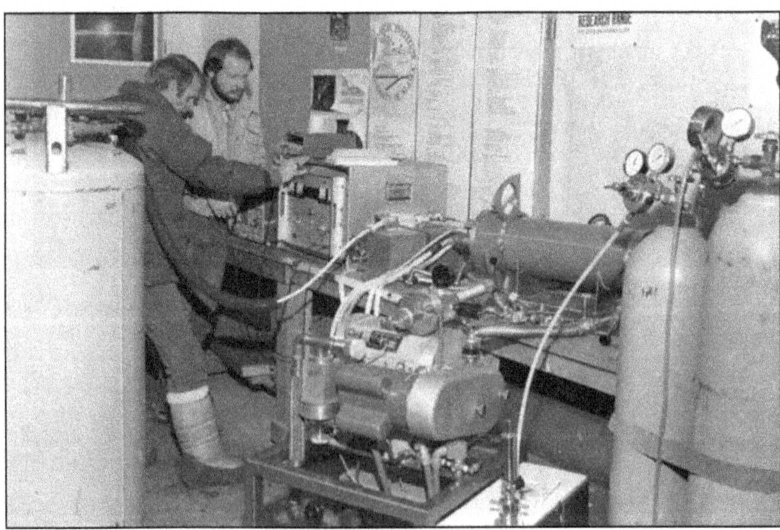

Technicians visiting Poker Flat from AFGL and USU assemble and test a sounding rocket payload in February 1980

An engineer assembles a sounding rocket for launch during the winter 1980 campaign at Poker Flat

First two stages loaded on Poker Flat launch pad 4 in February 1980. The insulated box around the second stage is not yet installed

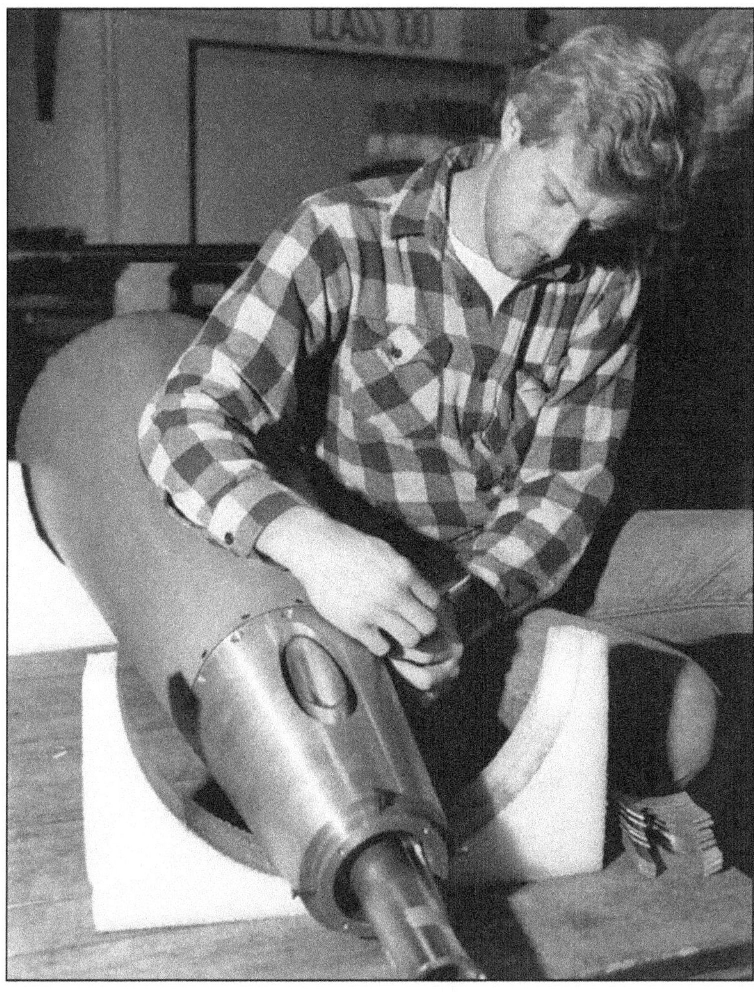

A technician assembles the nose cone of a sounding rocket in February 1980

Technicians from AFGL and USU analyze the nose cone of a sounding rocket in February 1980

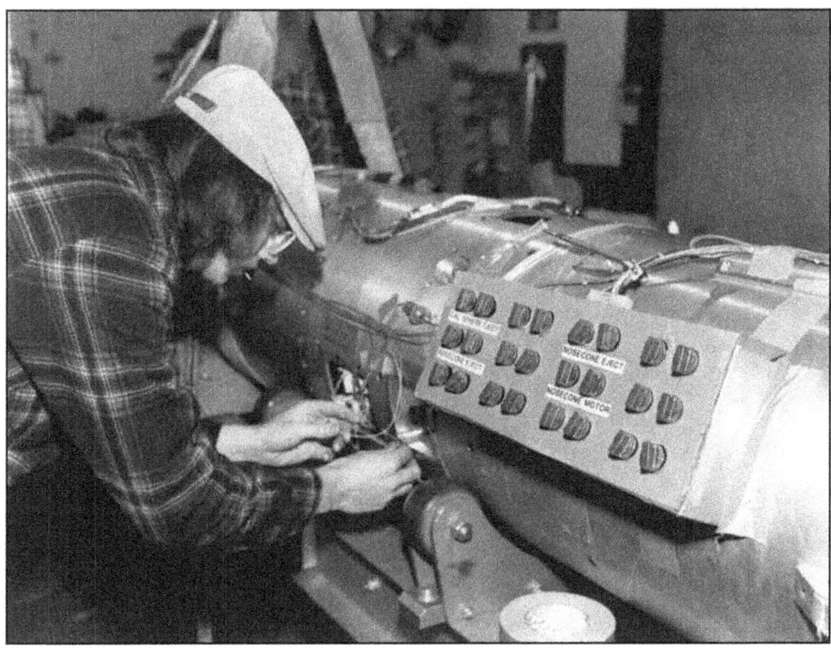

A technician from AFGL and USU performs diagnostic work on telemetry equipment housed in the payload of a sounding rocket before launch, February 1980

Left to right: Neal Brown, Carroll Coe, Tom Hallinan, and Eldon Thompson in the blockhouse during a rocket launch campaign in November 1974

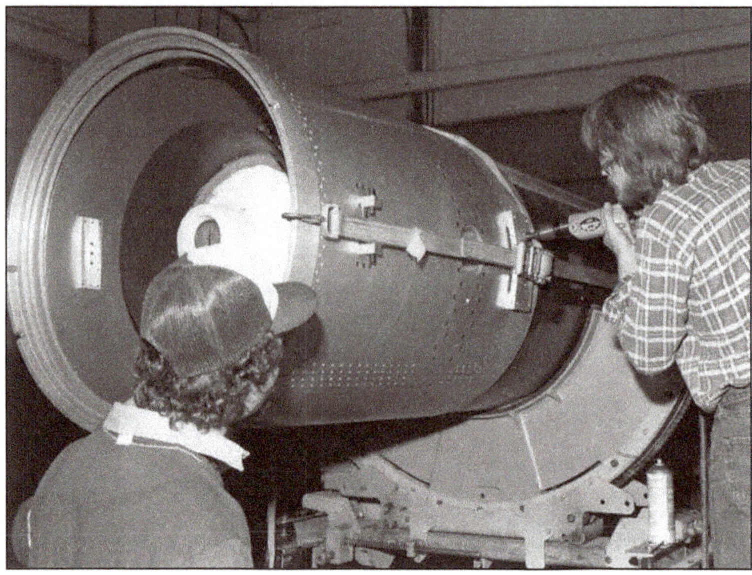

Engineers from AFGL and USU assemble a sounding rocket for launch during the winter 1980 campaign at Poker Flat

Terrier-Malemute sounding rocket loaded on launch pad 3 at Poker Flat in the winter of 1976

Poker Flat was a family operation. Here a portion of a rocket is being transported to the Payload Assembly Building at Poker Flat in the winter of 1990. Left to right: Gar Bering, Lars Osborne, and Daniel Osborne

By 1976, Poker Flat had built an optical observatory at the top of an adjacent hill to collect auroral data and make launch/no launch decisions

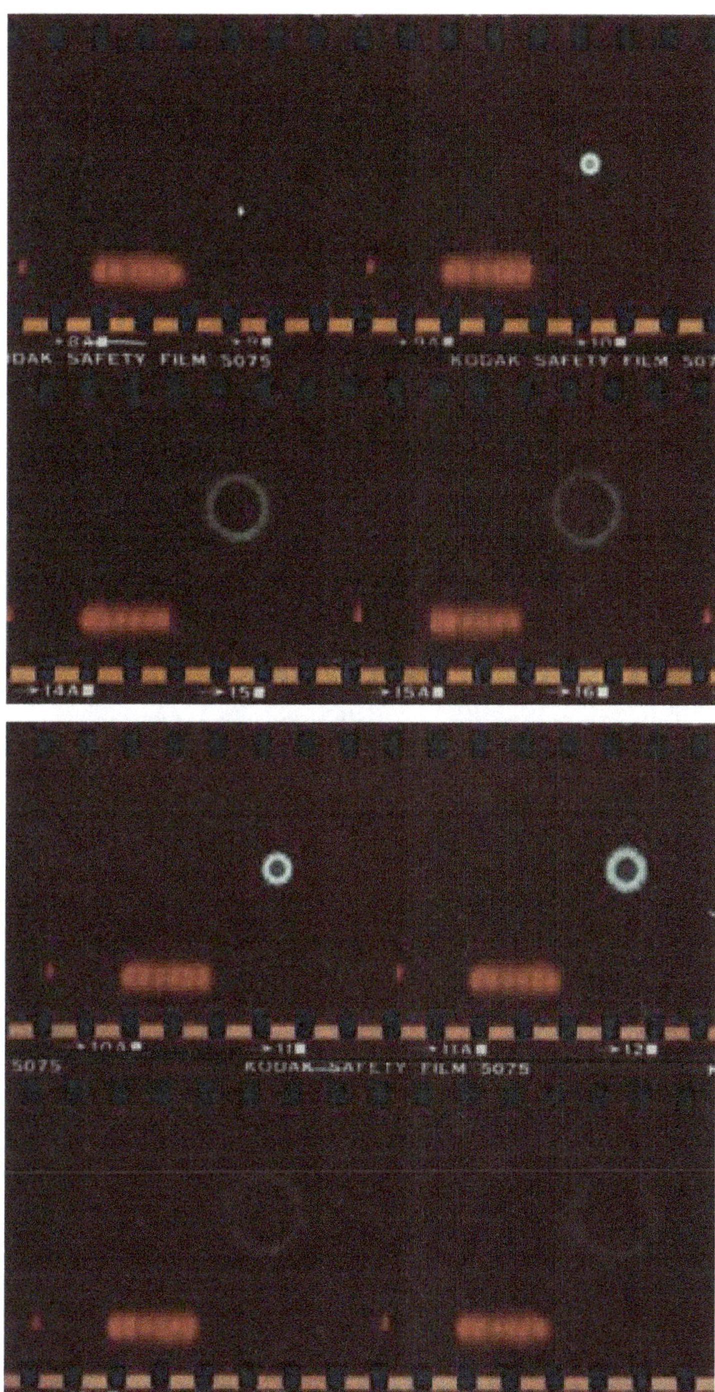

Bubble maker, a Gene Wescott experiment, was a Taurus-Tomahawk with a radial-shaped charge. It was launched on March 25, 1981. This series of shots show the barium release that resulted. Photo by D. Bostow

THE ALASKA LEGISLATURE

* HONORING *
* PROFESSOR NEAL BROWN *
* UNIVERSITY OF ALASKA FAIRBANKS *

The Nineteenth Alaska State Legislature takes great pride and pleasure in recognizing the achievements and contributions of Neal Boyd Brown, assistant professor of physics and former director of both the Poker Flat Research Range of the Geophysical Institute and the Alaska Space Academy at the University of Alaska Fairbanks. His retirement in June of 1995 after twenty-five years of service leaves a long list of professional accomplishments with the University.

Neal Brown first came to the University of Alaska Fairbanks as a graduate student in 1963. He received his master's of science in geophysics in 1966. After serving as a senior research assistant and assistant engineer at the Geophysical Institute, Mr. Brown became an assistant professor of geophysics in 1969, a post he continues to hold.

In 1971 Mr. Brown became supervisor of the Poker Flat Rocket Facility of the Geophysical Institute where he served until 1989. One of the nation's busiest space research facilities, Poker Flat is the world's only university-owned rocket range. Under Mr. Brown's direction, it led and still continues to lead the world in studies of the aurora borealis and other upper atmospheric phenomena.

Mr. Brown has helped to inform, enlighten, and entertain the world with his numerous scientific and general interest articles, reports, lectures, and publications. He has also made public appearances on national television shows including PBS's "Newton's Apple," the "Arctic Light" program on the Discovery Channel, and "Good Morning America." In addition, he helped to produce the video "The Aurora Explained."

Leaving his position as director of Poker Flat in 1989, Mr. Brown undertook a sabbatical at the Laboratory of Atmospheric and Space Physics at the University of Colorado. Since that time he has pursued long-standing interests in noctilucent clouds, high-time resolution, low-spectral resolution studies of the aurora, and science education.

In 1993 Mr. Brown conceived the idea for the Alaska Space Academy, a week-long summer camp for middle school students. In its first year, the camp attracted more than 110 students from Alaska, Canada, and Russia to attend the week-long sessions at Lost Lake, just south of Fairbanks. With adventurous, hands-on learning projects, the students were introduced to the scientific method using fun activities. Believing in the new program, Mr. Brown gave not only his time and energy but personal resources to the space camp.

We, the Nineteenth Alaska State Legislature, commend Neal Brown for his twenty-five years of excellent public service and wish him well in his retirement. We further hope that he will continue his long history of contributing to his community in the years to come.

SPEAKER OF THE HOUSE PRESIDENT OF THE SENATE

Date: March 1, 1995

Requested by: Representatives Davies, Brice, B. Davis, Elton, Foster, James, Kubina, Mackie, Navarre, Sanders, Vezey, Willis; Senators Frank, Miller, Sharp, Zharoff, Hoffman, Lincoln, Pearce, Duncan, Leman

Alaska Legislature honoring Professor Neal Brown, University of Alaska Fairbanks, March 1, 1995

ALASKA LEGISLATURE

IN MEMORIAM

NEAL BOYD BROWN

The members of the Thirty-Second Alaska State Legislature join family, friends, and community in honoring and remembering the life of Neal Boyd Brown, who passed away at the age of eighty-two on July 19, 2021.

Neal was born to Kenneth Wayne and Ruth Alvina (Boyd) Brown in Moscow, Idaho, on December 13, 1938. Growing up on the family farm in eastern Washington, Neal learned how to build and fix almost anything mechanical or electrical using any implement he could find. This skill helped him earn an Eagle Scout ranking in the Boy Scouts. Neal graduated from Pullman High School in 1957 and received a Bachelor of Science in Physics in 1961 from Washington State University. After experiencing the aurora borealis for the first time during a NASA research position in Thule, Greenland, he received a Master of Science in Geophysics in 1966 from the University of Alaska Fairbanks (UAF) and became an assistant professor at its Geophysical Institute.

Neal then spent eighteen years supervising and directing studies of the aurora at the Poker Flat Research Range - the world's first and only university-owned, scientific, rocket launching facility located approximately thirty miles north of Fairbanks. In 1985 and 1986, he collaborated with two other scientists, Tom Hallinan and Daniel Osborne, to create the Aurora Color Television Project, a first, broadcast-quality, color television footage of the aurora and one of his proudest accomplishments. He took a short sabbatical in 1989 at the Laboratory of Atmospheric and Space Physics at the University of Colorado where he developed a passion for public speaking about the aurora. In 1993, he conceived and developed the idea for the Alaska Space Academy, a week-long summer camp held at Lost Lake for under-represented, rural, and Indigenous middle school students from Alaska, Canada, and Russia.

Upon Neal's retirement from UAF in 1995, he was honored by the Alaska State Legislature for twenty-five years of service and, in 1997, received an honorary UAF Doctor of Laws. Retirement did not really stick, of course – Neal returned to UAF at their request from 2002-2008 to direct the Alaska Space Grant program and, in 2003, created his own consultancy, Alaska Science Explained, where he served as chief scientist and shared his love of the aurora, science, rockets, and space through teaching engagements with local schools, the Osher Lifelong Learning Institute, and other community groups throughout Alaska, New Hampshire, Vermont and more. One of his favorite venues for aurora lectures was Camp Denali in Denali National Park, where he could teach tourists about the Alaskan night sky amidst the beauty of the Alaska Range.

When not lecturing or educating about the aurora and Alaska, Neal spent time exploring genealogy and operating his HAM radio. He especially loved to spend time with his wife, Fran Tannian, and often surprised her with bouquets of fresh flowers. They enjoyed traveling together, birdwatching, and dancing with their Bichon Frise, Molly. Neal stayed in close touch with his beloved children, grandchildren, and extended family through countless letters and other forms of communication.

Neal is survived by his wife of forty years, Fran Tannian; children, Kristopher David Brown (Rachel), Melody Brown Burkins (Derek), and Nathaniel Scott Brown (Tina); stepsons, Steven Ross Sweet (Iva) and Michael Scott Sweet (Laura); and grandchildren, Ari Morris Brown, Noa Brown, Riley Logan Burkins, Porter Brown Burkins, Zacharias Tsiakalis-Brown, and Michael Tsiakalis-Brown.

The members of the Thirty-Second Alaska State Legislature extend their condolences to the family, friends, students, and community of Neal Boyd Brown. Neal was a pioneer, and his passion and drive will be greatly missed.

LOUISE STUTES
SPEAKER OF THE HOUSE

PETER MICCICHE
PRESIDENT OF THE SENATE

SEN. SCOTT KAWASAKI
SPONSOR

REP. GRIER HOPKINS
CO-SPONSOR

REP. ADAM WOOL
CO-SPONSOR

Date: March 28, 2022

Alaska Legislature in Memoriam, Neal Boyd Brown, March 28, 2022

University of Alaska Fairbanks

The Board of Regents of the University of Alaska System
on recommendation of the University Faculty and by virtue of the
Authority vested in Them by Law have conferred upon

Neal Boyd Brown

the degree of

Doctor of Laws

for his commitment to science education and his leadership in creating
lasting programs to enrich the lives of Alaska's youth
with all the Rights, Privileges, Honors, and Obligations pertaining thereto
Given at the University of Alaska Fairbanks, this month of May, A.D., 1997.

University of Alaska Fairbanks Doctor of Laws degree awarded to Neal Boyd Brown, May 1997

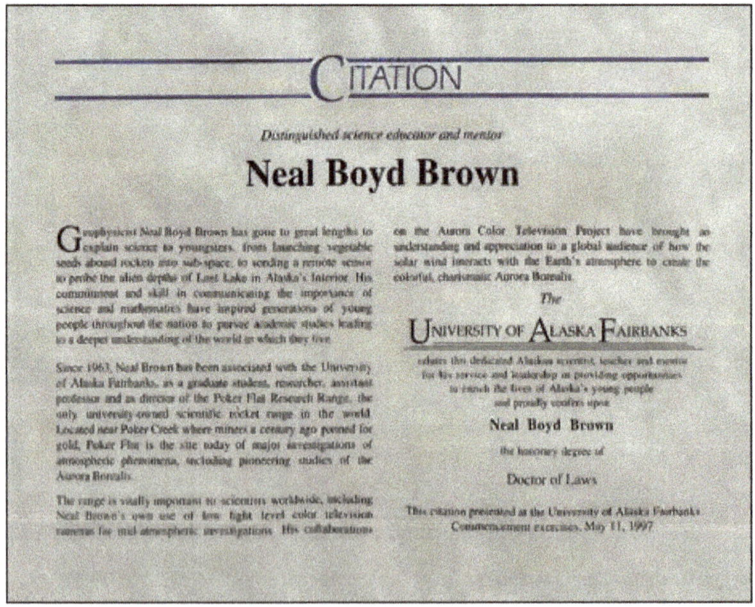

CITATION

Distinguished science educator and mentor

Neal Boyd Brown

Geophysicist Neal Boyd Brown has gone to great lengths to explain science to youngsters, from launching vegetable seeds aboard rockets into sub-space, to sending a remote sensor to probe the alien depths of Lost Lake in Alaska's Interior. His commitment and skill in communicating the importance of science and mathematics have inspired generations of young people throughout the nation to pursue academic studies leading to a deeper understanding of the world in which they live.

Since 1963, Neal Brown has been associated with the University of Alaska Fairbanks, as a graduate student, researcher, assistant professor and as director of the Poker Flat Research Range, the only university-owned scientific rocket range in the world. Located near Poker Creek where miners a century ago panned for gold, Poker Flat is the site today of major investigations of atmospheric phenomena, including pioneering studies of the Aurora Borealis.

The range is vitally important to scientists worldwide, including Neal Brown's own use of low light level color television cameras for mid atmospheric investigations. His collaborations on the Aurora Color Television Project have brought an understanding and appreciation to a global audience of how the solar wind interacts with the Earth's atmosphere to create the colorful, charismatic Aurora Borealis.

The

UNIVERSITY OF ALASKA FAIRBANKS

salutes this dedicated Alaskan scientist, teacher and mentor
for his service and leadership in providing opportunities
to enrich the lives of Alaska's young people
and proudly confers upon

Neal Boyd Brown

the honorary degree of

Doctor of Laws

This citation presented at the University of Alaska Fairbanks
Commencement exercises, May 11, 1997

Citation honoring distinguished science educator and mentor Neal Boyd Brown from the University of Alaska Fairbanks, May 11, 1997

Neal Brown took this photo on June 16, 1983. He considered it the best photo he ever took of a rocket lifting off at Poker Flat. He wrote, "It has been used in many professional science and educational publications and hung in my office at Poker Flat for many years"

Neal Brown poses "conducting" the aurora, 2015. Photo by Marketa S. Murray

Chapter 5
Poker Flat Infrastructure

Improving Range Safety

Everyone who works at Poker Flat Research Range is trained in safety, particularly with regard to working around and launching rockets and recovery from an accident. But sometimes people who were not regular workers were unaware of our procedures. We had a handful of incidents that made me aware of the need to improve safety procedures at the range.

When I came to Poker Flat in August of 1971, the entrance road off the Steese Highway was unmarked and ran within 100 feet of pad 1. A padlocked gate was hung from a pair of 6-inch-diameter, 12-foot-tall steel posts embedded in concrete. We left the gate open except when we went home at night or if we were counting down a rocket and sent the roadblocks out to block the highway from the Chatanika Lodge to a mile beyond Poker Flat on the Steese. A World War II field phone was strapped to one of the poles, connected to another field phone in the blockhouse, though I don't recall anyone using it.

In 1973, we built a new road into Poker Flat with a lighted entrance. We installed a microphone/speaker so we could talk to whoever showed up at the gate, and we had a television surveillance camera next to it. We used the same setup for a lockable gate.

One day in 1973, an 18-wheeler brought a load to Poker Flat on a 40-foot flatbed. We unloaded the trailer by late afternoon, but we did not notice that the 18-wheeler had not left the range. We were counting down a rocket and were at a relaxed T-minus 20 min and holding. Our wind-weighter Merritt Helfferich gave us an update to swing the launcher east. Suddenly we saw the idling 18-wheeler, puffing exhaust from the twin stacks behind the cab, the trucker asleep in the extended cab. We woke him up and had him move his truck to another area more than 100 yards from the launcher, which was considered safe if he was in the cab.

N. Brown, *Northern Lights and Rocket Flights*, Springer Biographies,
https://doi.org/10.1007/978-3-032-14598-7_5

A few weeks later, we saw something amiss on the television cameras aimed at the launchpad: people moving around looking at the launcher with its rockets and payload. This time it was a well-meaning geology professor I had worked with many times, leading a field class, and he stopped in to show them the rocket. Fortunately, again we were at T-minus 20 min and holding. The next day I created several posters for the Geophysical Institute and a handout for every member of the GI faculty and staff, asking them to please call before visiting Poker Flat.

Radio-Controlled Gates to the Launchpads

For additional security, I asked the head of our machine shop, Larry Kozycki, to design and build two electric motor-driven arms that we could install across the two roads leading to the launchpads and control with an electric switch from the blockhouse. I assumed we would have to wait until summer and pour a concrete pad across the road with the motorized arm on one side and something on the other side to rest the arm in the down position, where it would be about headlight-high for a car or truck. Larry gave me the design in the same meeting as I brought up our need. He was a genius. He soon made a simple triangular stand for the electric motor mechanism and chose a garage door opener with a remote radio control holding a piece of plastic about halfway across the road, painted it yellow, and put a spiraling black stripe along the length of the arm. This was in place from about 1975 through 1993. It was great that we did not have to run electrical wiring to these units, just power from our nearby headbolt heater outlets.

In 1993 a radio-controlled front gate was added. When the range office saw someone arrive on television monitors, they would open the gate and tell them to come check in at the range office, which was just out of sight on the curving entrance road. Staff could also open the gate with a card issued in their name. Each time they entered, it logged them into a computer on range and kept a record of their coming and going.

Moose on the Launcher

In October 1971, a group of site selection advisory people, including government and university types, had come to Poker Flat over the weekend. Neil Davis was giving them a tour while expounding on how Poker Flat was no longer a bush-league organization. As Neil led them over to the rocket launchers, they all started giggling. When Neil looked up, he saw a freshly gutted moose hanging from one of the rocket launchers. Seems some of the crew had gone moose hunting earlier in the week and had had trouble with dogs trying to get at the carcass, so they hung it there for safe keeping.

Electrical Grounding

When working on a radio chassis, you should always make sure you have a single point where all ground leads are put together. If you don't, noise can get into the system. This is similar to having a ground pin in all of the electrical outlets in your home. Having these grounded keeps noise out of those circuits, but, more important, if you have a 120-volt alternating current wire reversed, it will trip a ground-fault circuit interrupter. At Poker Flat, with its buildings tens of yards or even miles apart, this does not work. The ground under the blockhouse was connected to the telemetry building through groundwater, creating an unintended electrical ground. We troubleshot a lot of problems through the years, but the problems finally disappeared only when fiber optic cable was installed to carry communication signals throughout Poker Flat.

Really good low-resistance grounding is super important when you are handling rocket motors or working on payloads, especially when you are connecting two items. You don't want sparks flying around explosives. Heavy copper straps line the walls of the buildings where you do this work, with large alligator clips to attach to the item you are working on. In addition, when building up the rockets in the rocket assembly building or on the launchpad, you connect yourself via these alligator clips to a nylon suit big enough to cover your cold-weather clothing.

Liquid Coolants for Infrared Detectors

In 1973 we had our first experience with liquid nitrogen as a coolant for near-infrared sensors aboard payloads. By 1975 we were also dealing with liquid hydrogen. In both cases we had to store the canisters outside and make sure they did not plug up with moisture from the air. Even with 5% relative humidity, we had to brush frost off several times a day. Hydrogen was the more worrisome, as the temperature inside was at least four times colder than liquid nitrogen.

Chatanika Incoherent Scatter Radar

In 1971, Poker Flat was to provide infrastructure for the Chatanika incoherent scatter radar (ISR), a powerful radar transmitter used to measure the properties of the ionosphere and the upper atmosphere. The Department of Defense had moved it from Menlo Park, California, to 28-mile Steese Highway, two miles short of Poker Flat. At first we just provided things like power and communication for the facility. The ISR was theoretically proven to work, but no one had ever completed an experiment that verified it worked. The Chatanika ISR energized a huge volume of the ionosphere. As the ions relaxed, they gave off a very weak signal. The first experiments to prove the ISR worked were done with correlated data from sounding rocket payloads launched when the ISR was working. These proved the concept to be theoretically and experimentally valid and led to further improvements in the software so that they could do Doppler shift measurements of the ion/electrons to determine ionospheric wind speeds and directions. Later the Department of Defense moved the ISR to Sondrestrom Fjord, Greenland, where it sits inside the auroral zone.

Electricity Demand Costs

When I came to Poker Flat in August of 1971, Eldon Thompson and the crew were installing 110-kW heaters on the tops of launchpads 1 and 2. I asked Golden Valley Electric Association (GVEA) in Fairbanks if we would be hit with extra charges if we exceeded a certain level of use. I knew that all of the wind tunnels at NASA Ames were electrically powered, and if they exceeded local supply, it could cause circuit breakers to trip in San Francisco or Oakland, 40 miles away, and could even cause transformers to fail because of the sudden load drop.

GVEA said they had measuring equipment on the lines that fed electricity from the coal-powered steam plant in Fairbanks. The army post at Fort Wainwright, the University of Alaska, and Eielson Air Force Base all had their own coal-fired plants and could share with everyone in case of an emergency. GVEA said they would watch our particular line as it left Fairbanks, because we were the prime user, and call us if there was a problem. We never had a problem.

When the Chatanika ISR was installed, monitoring equipment was also installed at the substation that fed it, Poker Flat, and the hilltop area where the army Meteorological Rocket Network was set up. So far so good, until we built the Poker Flat optics site in 1974. GVEA asked us to monitor our power usage. They would do the same, and we would work cooperatively to evaluate how to proceed in the future. I hired Brett Delana at the Geophysical Institute to do our portion of the work. Brett got records for winter months, when our load demand on GVEA was high, and summer months, when it was low.

When the rather large Davis Science Center was built on top of the hill at Poker Flat with completely electric heat, the situation was renegotiated. We never exceeded the capability of the transformers or lines delivering power to Poker Flat, although I think GVEA did upgrade the lines from Pedro Dome to Poker.

Meteorological Tower

In 1971, we had a 30-foot-tall meteorological tower with two wind speed and direction instruments. The data from these instruments were used to calculate launcher settings for the unguided rockets we launched. When I asked how the wind speed and direction information was used, Neil Davis, Eldon Thompson, launch officer

Carroll Coe, and Poker Flat's wind-weighter Jim Woolf gave me a primer on how the sounding rockets launched from Poker Flat were much like small model rockets, except they dive into winds, not drift with them. But the general idea is that rockets are sensitive to local winds, which can cause them to fly and land where you might not want them to.

A launch countdown began roughly three hours in advance of a rocket launch with the launch of a standard weather balloon. NASA's VERLORT S-band radar was used to track its flight. A radar-reflective bowtie shape made of aluminum foil was hung beneath the balloon. Radar track data was used to provide wind speed and direction information as a function of altitude from the surface to near 100,000 feet, and this was done at least once every 4 h of a rocket launch window. In addition, roughly every 20 min a 3-foot-diameter helium-filled balloon was also launched and tracked. Its radar track was used to provide updates on the wind speeds and directions below 10,000 feet in altitude. However, Jim stressed that data on the most important and likely the most changeable winds affecting a rocket's flight came from the instruments on the meteorological tower, and we needed a taller tower to measure winds higher up.

When I asked how much change one could safely make in azimuth and elevation to get a sounding rocket to land in a safe place, I learned a bit more about how such decisions were specific to a given configuration of the rocket motors stacked one above the other to carry a scientific payload into the aurora. I got the general idea that only a few degrees offset in elevation and a few tens of degrees in azimuth were acceptable. I saw how geared electric motors in the launchers could be controlled and monitored from inside the blockhouse to make these elevation and azimuth changes and that you never pointed a rocket so that if all the winds suddenly died at launch it would fly and land where you did not want it to.

Neil, Carroll, and Jim all said they wished we had a taller and better-equipped meteorological tower. Eldon Thompson told me that the Geophysical Institute owned enough material that we could easily put up a 265-foot-tall tower at Poker Flat. Eldon already had six 10-foot-long sections of new tower, and he told me the material was identical to two 100-foot-tall GI radio towers at the Ballaine Lake field site near campus. These towers were no longer used or needed, so they were ours for the taking. There was this one little problem: the two towers, which were connected at the top with a 50-foot crossbeam, had concrete slabs underneath them that had jacked up and down during freeze and thaw cycles so that the towers were now slightly tilted toward one another. Eldon said the towers were not damaged, but the crossbeam was under pressure. He said it would be best to hire a local company with steel workers to take down the towers, because the stress they were under would make it dangerous for two people to work on the two towers. If they let the strain go off too quickly, the towers might throw the person off the tower even if they wore safety belts.

We contacted a local company that specialized in such work and they gave us a quote of $25,000 to take the two towers down. Part of the reason the cost was so high was that work at this height required members of the local steelworkers union to be involved.

We had not budgeted any funding for this project we were dreaming up, so we looked for an alternative solution. We reasoned that our range users would want us to have a taller, better-equipped tower but figured that writing a proposal and getting it funded was going to take a least a year, and we were already hoping that we might get these towers down and a new tower up by the start of the March 1972 launch programs.

Eldon contacted Jack Smith and Eric Forrer, both former Poker Flat employees, about taking the towers down. Neither knew anything about climbing towers or taking towers apart, but they needed money and Eldon thought they were cool-headed enough that they would not get hurt. He said he would talk them through how to do it and that he had a gin pole he would loan them. A gin pole is an upright mast that is guyed at the top to maintain it in a vertical or nearly vertical position and is equipped with suitable hoisting tackle.

After they won the one-time bid for $2000, Eric and Jack went down to Samson's Hardware and bought 500 feet of half-inch nylon and two lineman's safety belts with back pads and metal side clips and giant buckles. It took several days to get everything into position, double-check, and think through what they were going to be doing, because of course everything was almost a hundred feet off the ground. Eric and Jack would take the tower sections down and load them onto a trailer that Eldon would leave at the Ballaine Lake site. Eldon would then periodically stop by with his four-wheel-drive Dodge Power Wagon truck and pull the trailer loaded with tower sections out to Poker Flat, unload the sections, and bring the trailer back to Ballaine Lake to load again.

Eldon also told me he had rigged an automotive handyman jack that would be extended and secured from one tower to the crossbeam. Using the handyman jack, they could change the tension between the two towers to lessen the chance of the tower flipping Eric and Jack off. There was a bit of a horizontal strain being taken by the antenna, but it didn't amount to more than a few inches and they just loosened the clamps.

One of the most exciting moments was when Eldon came wheeling into the yard in his truck and literally ran up one of the towers in street shoes to show us how to rappel down the line to save time. He seemed absolutely confident and relaxed a hundred feet above the ground, and we were completely petrified because we thought every move he made would be his last. Finally, he just soared off down the line, kicking himself away from the tower, unhooked, and drove off.

The object on top was a big antenna lying horizontally across the top of the two towers, and it wasn't immediately apparent how to get it down because all the guy wires were in the way. The guy wires were not precisely at the top but maybe one 10-foot section down. So the first move was to remove all the guy wires that were in the way, leaving the antenna on top of more or less unsupported top sections. One guy wire on each tower was parallel in spacing and angle to a similar guy wire on the other tower. Eric and Jack used these two guy wires as a ramp to slide the antenna down. Each man was on top of his tower with a rope tied to the antenna. They kept the antenna horizontal as they lowered it down to the ground.

The real challenge was to keep everything that they did to the antenna symmetrical so it stayed horizontal. Eric and Jack had to be on top of the towers to make the last disconnect and set the antenna on the parallel guy wires. They each had to put the main antenna pole on their shoulders and lift it over projecting pieces of the top section. Their apprehension was pretty high because they were not used to such dizzying heights and because they were afraid that something one of them did might throw off or jeopardize the other man.

About the time Eric and Jack got the antenna down and were celebrating being back on Mother Earth, Eldon showed up and declared a management crisis. The companies that had lost the bid were raising hell about the technicalities of it all, and Eric and Jack had to go down and buy a business license. But then the United Steelworkers union also got into the act because they were of course not in the union. The union sent a guy out to the site who sat in his car for hour after hour, watching the operation and generally serving as a significant distraction. In the end he stopped showing up without ever having said a word.

Once the antenna was down, the towers were not really connected anymore, so Eric and Jack took off all the guy wires and used the gin pole to lower one section at a time. It was much easier for one of them to be on the ground with control of the line and the other to unbolt a section and free it up for the drop. There wasn't much excitement after they got the hang of the procedure.

A few weeks later, with 200 feet of the new meteorological tower in place and guyed, Merritt Helfferich and I watched as Eldon and Beau Battey hung in their safety belts 65 feet above the blockhouse on the meteorological tower. A helicopter was about to lift the last sections of tower. By radio Eldon told the helicopter pilot and his crew they were ready. The helicopter maneuvered almost straight up while taking up slack in the cable attached to the top 60 feet of preassembled meteorological tower that lay on the ground. It hovered momentarily when it had taken all the slack out of the line and then maneuvered until the cable was straight up and down. Then they began lifting the 60 feet of tower until it stood straight up and down and was free of the ground. Now, cautiously, they moved forward, gaining altitude. They flew a circle and came back with the bottom of the tower section hanging straight and barely above the top of the tower Eldon and Beau were on.

Eldon and Beau tried to grab the rope attached to the tower section hanging over their heads, but it blew out of reach in the helicopter's downdraft and wrapped around one of the guy wires that held the section of tower that Eldon and Beau were on. The rope slipped up the guy wire and snagged tight where the wire met the tower, just below Eldon and Beau. The rope, the piece of tower above, and the helicopter above them no longer formed a straight up-and-down line. The helicopter maneuvered to regain vertical alignment, but clearly things were going bad fast.

We watched Beau undo his safety belt and move as fast as he could down the tower. As I watched the helicopter straining and maneuvering, I thought to myself, "If the pilot gets in any more trouble, he is going to release that tower to save his helicopter and crew. If he does that, the chances are high that the tower Eldon and Beau are on will be dragged down by the falling tower. They will likely be killed."

Beau stopped. We could see his arm move, and the rope came undone.

Now that the helicopter was free, the pilot was able to regain control. He flew in another slow circle, and as he came back, he laid the tower down as carefully as he had picked it up.

Eldon and Beau had meanwhile come down the tower. Eldon held his hand to his neck and was clearly in pain. The rope leading from the tower to the straining helicopter had been against Eldon's bare neck, trapping him to the tower in his safety belt. When Beau cut the half-inch rope loose, it ran quickly up and away alongside Eldon's neck, giving him a deep rope burn about 12 inches long on his neck. It was horrible to look at and took months to heal. That was the end of that workday for everyone involved.

After the helicopter fiasco, Eldon talked to Jerry Duncan, who was working at Poker Flat in the fall of 1971. He had never been up a tower before but was game to do the work. We paid Jerry a laborer's wage and added an incentive that he could stay in a room at Poker Inn for free. Eldon taught Jerry what he needed to know. In not terribly pleasant weather, the crew at Poker helped Jerry, and each day he went up the tower by himself and got another piece into place. I'm pretty sure he finished before Christmas 1971, and that included running cabling up the tower and connecting wind-speed and direction sets.

Since the tower was over 200 feet tall, the FAA required that we put professionally built red blinking warning lights on top of it. We bought a unit that stood about 3-feet-tall and had two expensive rough-service light bulbs, for redundancy. The unit that caused the lights to blink was mounted in the blockhouse and made a reassuring rhythmic click, except when both lights had burned out. During nonlaunch season, when we worked days and not nights, the silence was sometimes our first clue that the tower lights were not working.

The voltage everywhere on range was 130 volts or higher instead of 120 volts, because the military-surplus transformers that connected Poker Flat to the GVEA transmission lines were slightly mismatched. The higher voltage caused at least one of the two tower light bulbs to burn out each month. We eventually discovered that we could put a resistor in line with the electrical power in the blockhouse, and then the blinker and lights would continue to function.

Home on the Range

by Wayne Pinger

Our family of four, plus two dogs, two cats, and one box turtle, arrived in Fairbanks in the fall of 1972 just as the snow and cold weather were starting. I had been hired by the Geophysical Institute to do support work for two University of Alaska scientists who were studying the northern lights at the Chatanika Radar, a facility funded by both NASA and DNA and managed and operated by Stanford Research Institute. The radar site was about three-quarters of a mile from the Poker Flat Research Range, just down the Steese Highway.

I had applied for work as an electronics technician in 1970 when first visiting Fairbanks and had waited not-so-patiently for over 2 years for a position to open. So, just 4 weeks after accepting the job, our four-vehicle caravan arrived in Fairbanks, we moved into university subsidized housing at Yak Estates, took several big breaths to relax, and assumed the roles of tenderfoot Cheechakos.

Funding had not yet arrived for the project. When it did, I would be assisting Skip Bates and Bob Hunsucker to operate the radar facility, record, and later disseminate the radar data. Also I was to operate a series of all-sky cameras and help when I could with any other work that was needed at the radar site. Some of these tasks, I learned later, proved to be unique and could only be described as "Typical Alaska Operations," though at the time I was totally unaware of the meaning of that phrase.

With the funding for my position at the radar site still not available, Poker Flat Research Range agreed to employ my services temporarily. The range was expanding and there was a need for some technical help for wiring, welding, and other things. That was when the adventures began.

I arrived at the range one cool plus-10-degree Friday afternoon and met the range manager, Neal Brown. He introduced me to the range foreman, Eldon Thompson. Eldon drove me on a tour of the range while dropping Neal off at a newer building, the rocket storage building. We continued to another building where I was introduced to the crew of eight, "the Poker Flat Irregulars." They were having coffee in what seemed a lunch room in one of the bigger central buildings called Poker Inn. The group consisted of some men, one lady, and one person with quite long hair, beautiful fingernails, but a baritone voice. I was unable to tell which sex, if any, he or she was. In retrospect, it might have been one of the more interesting groups I had ever met.

Neal arrived and had a few words with Eldon as I was introducing myself and shaking hands with everyone. Then Neal left and I seemed to be in the care of Eldon, who everyone called ET. I was waiting for some sort of work assignment when ET mentioned an initiation of some sort for new crew members, and mine, since I was temporary and an electronics technician, would be replacing a burnt-out bulb on a nearby light standard.

On the top of one of the many nearby pickup trucks parked at Poker Inn I had seen a fairly large three-legged ladder and somehow I put this together with the mentioned task. So when asked if I had a fear of heights I said, "Not really." A bit of a fib, I guess; though I don't have great fear, I am never very comfortable more than 8 or 10 feet off the ground unless I'm surrounded by wings. But the ladder seemed not all that tall, so I said nothing. Then I was given the box with the light bulb in it, and it proved to be no ordinary bulb because it was red and about a foot tall.

After a cup of coffee, we drove down a road in what was called the auger-truck, which proved to be a military surplus World War II six-by-six converted to a flatbed. The truck had a large power take-off driven winch mounted on its front and an auger, for boring holes I guessed, mounted on the back. We drove a short way to a multiguyed tower that had a red lamp on top that was now blinking. I slowly realized that the light bulb I was holding was to somehow be placed on the top of that tower. I was told there were two bulbs on the tower: one was always active and the other was a spare that would faithfully blink when the main bulb ended its boring but necessary life as a signal to low-flying aircraft.

The tower was ungodly tall; maybe about 150 feet and guyed at three separate levels. High enough, I was told, that no one usually climbed it but instead people were winched up to the top in a bosun's chair. Today the lucky traveler—that would be me—would be hauled up by the auger truck's front-mounted winch, which would pull a very heavy nylon climbing rope through a snatch-block at the bottom of the tower and a simple single pully at the top; this contrivance would elevate a fearless climber to new heights.

I was told by several of the Irregulars, who all seemed to be talking at once, that the view would be worthwhile and to be on the lookout for any moose with horns. Surely, they meant antlers? They said the itinerant moose or mooses might be on the far side of the river that I was told was just across the road; wherever the hell the road was. Somebody said Richard had already shot one moose but thought there was at least one other around.

But quite selfishly I was thinking less of moose and more of my family. I was wondering if there were any insurance policies for U of A staff; surely there were some. That thought left me when I noticed ET seemed to have a devious smile, but I ignored that, thinking instead how my son would likely avenge my demise in future years when he became an adult.

Many hands were now helping me into the bosun's chair, and I thought maybe I had to pee but wondered if that might exhibit some fear. Stubbornly I decided to just hold it.

Suddenly, and maybe not all that surprisingly, I was feeling unwell. Common sense told me to back out of this situation, but my ego was saying, "Don't let these guys disparage the new kid from California." Unfortunately, my ego won the contest, so after being fitted out with both a climbing harness and a hard hat, and with toolbox in one hand and the rather large red bulb in a small backpack, I was slowly being winched up the tower.

At about 10 feet I thought, "This is okay." At 20, I thought maybe it was a bit of a mistake, and at 30 or so, I was beginning to see just how big the mistake was. Thirty feet seemed like half a mile as I looked down with zero interest in finding the road or the river or a moose. I was about to scream something when the ride to the heavens stopped and I was slowly, but in small jerks, lowered to the ground.

Though it was well below freezing, I was sweating. The crew was laughing but not so much at me, more like with me. It seemed I was now a member, a new one of course, of this group of Poker Flat Irregulars. Some of them had gotten the same ride, and it had ended the same way because for insurance reasons, ET was the only one to replace the FAA-required red blinking lights on top of the tower.

It was a good laugh that ended when ET donned the harness and hard hat. With the toolbox in hand and the spare bulb in the backpack, he was winched to the top in what seemed to be about a 2-min ride.

On ET's ride up, I thought maybe the winch gear-box sounded a little noisy with some cracking sounds. I could just barely hear it over the auger truck's engine noise but being the new kid on the block and not all that knowledgeable about winches anyway, I said nothing.

I guess my arrival had slowed down the work a little and the sun was just about to set over the surrounding hills to the southwest. It was September, getting cooler, and the sky was slowly clouding up and snow was starting to fall. It was one of the first snowfalls of the season I was told, and there was a northly wind starting to come down the canyon from Cleary Summit to the south. Winter was on its way, and I was naïve enough at the time to be looking forward to it.

ET had on just a windbreaker and thin gloves and carried a handheld radio. After replacing the burnt-out bulb and making sure it worked, he radioed that he was cold but still okay. He wanted to come down now. It was then that we discovered that the winch would not properly release and play out the line.

The winch was driven by a power take-off and was designed to only pull. There was a brake lever to hold the cable when the power take-off was put in neutral, but there was no way to back up the winch mechanically. When released the cable was free, but it had to be played out by hand.

And the cable would have been free, except that one of the Irregulars had shot a moose a week earlier and had used the truck to retrieve the kill, which unfortunately had been on the far side of the Chatanika River. The river was low this time of year, but when the truck was crossing at what was assumed to be a shallow spot, the winch had been submerged for a short time but still long enough for the gearbox on the winch to flood. Though frozen, the winch had no problem pulling as the strong gears easily crushed any ice trying to impede the mechanism, but the ice was still strong enough to impede the winch when playing out the climbing rope to allow ET back down from the tower.

The sun had set and it was starting to be a bit windy up there. ET was on the radio saying he was very cold and losing feeling in his fingers enough that he felt incapable of free-climbing down the tower. One of the guys pulled a trailer-mounted portable heater up to the front of the truck. It started quickly and blasted the winch and, unfortunately, also the front of the truck, with several million BTUs of really hot air. The heater is called a Herman-Nelson and is one hell of a heater. In a very short time for us on the ground, but a much longer time for our foreman at the top of the tower, the ice in the winch gearbox melted, along with the entire back of my Eddie Bauer rip-stop, goose-down-filled parka. My parka, or now half a parka, had been a parting gift from some friends in California, which I was now suddenly starting to miss. ET was ceremoniously but quickly lowered to the ground and boy, was he ever cold.

That should have been the end of the adventure, but the heater was left on as we spent some time in the warm, goose-down-filled blowing air, congratulating ourselves on ET's rescue. In just a few minutes, he was recovering nicely and getting some feeling back in his hands. We, the rescuers, also stood in the hot air because it was now nearly zero degrees.

ET was told the story of the iced-up winch. Just as the successful moose hunter was about to get his ass chewed, the radiator on the auger truck melted down. Steam and several gallons of blood-red antifreeze went up like a volcano, and the Herman-Nelson heater also melted the nylon climbing rope still attached to the winch. As the line melted through, the bosun's chair fell the final three feet to the ground, followed by about 300 feet of climbing rope that no longer stood ready to lift intrepid climbers to the top of the tower. ET was far from pleased because it meant sometime, in warmer weather, he would have to climb the tower the hard way to rethread the system.

The heater was quickly returned to wherever heaters are returned to, and two of us with knives cut, scraped, and pulled out of the winch what melted nylon line was left. We found a tarp and covered the front of the auger truck to keep it free of snow that by then had quit falling anyway. The job of radiator repair was left to the range mechanic, and he was also not pleased because that job would have to be completed right where it was and not in the heated high-bay where most of his work was usually done.

ET left the scene for a short period but soon arrived back. He was no longer antifreeze stained but was now properly dressed for the cold, and I think likely refueled with a wee dram or two of scotch. By then the mess was mostly cleaned up, except for the antifreeze from the blown radiator that was red and very apparent in the snow. ET was again about to vent his anger on the intrepid hunter when range manager Neal Brown drove up and asked, "Did you get the bulb replaced okay?"

"Yes we did, but we spilled a little antifreeze in the process," said ET.

Neal just nodded and said: "Let's be careful; I don't want this place to be a Superfund cleanup site in 20 years. Let's break for some coffee in the lunchroom. I want to tell everyone about some of the upcoming rocket launches that are planned by the folks from Wallops Island. They will be here in three weeks and we have some work to do. Also, damnit, and I'm only going to say this once: there is a moose hanging in the new Rocket Storage Building; that is un-good and I want it gone before morning."

Each summer we took the wind instruments off the tower to protect them from lightning. After Jerry left, Eldon devised a new way to more quickly service the tower. He installed a two-pulley unit just under the top of the tower and then ran a heavy 1-inch-diameter rope up and down the tower. To one end he attached a simple wooden board seat, and through another pulley at the base of the tower, he ran this rope out to attach it to our road grader. Beau and others went up and down the 265-foot-tall tower many times to service wind-speed and direction instruments using this system.

By summer 1972, we had our new meteorological tower in place and instrumented with wind-direction and wind-speed instruments at the 40-foot, 150-foot, and 260-foot heights. Six sets of wind instruments on the tower provided data to help determine launcher settings for a given rocket.

Neal Brown climbing the tower, located near the blockhouse and launch pads at Poker Flat, to inspect windset systems during the summer of 1978

The blockhouse in 1977 at Poker Flat with the observation tower used to monitor wind speed and direction behind it. The blockhouse, with a 3-foot reinforced concrete roof, was covered with an additional 4 ft of dirt to protect the crew inside in case of an accident during a launch. The blockhouse is where the command and control systems for launches are maintained

Handling Gas Bottles

The Geophysical Institute had built and deployed a balloon inflation launch structure on campus and at Fort Yukon to support the launch of instruments to study the atmosphere and electric fields created by aurora at 18.5-mile altitudes. The GI had gotten the research and development part done under an NSF grant, but due to the ruckus following the 1960 shooting down of American pilot Gary Powers in a U2 spy plane over Russia, Russia did not permit any overflights, especially by high-altitude balloons carrying any instruments. The GI program died. In the summer of 1971, Larry Sweet had the crew at Poker Flat scrape off an area beside the road to the substation and hired a contractor to move the balloon inflation building from University of Alaska to Poker Flat.

Larry also had a chain-link enclosure made for the gas bottles used to inflate the balloons. Range users also used gaseous molecular nitrogen for the attitude-control systems aboard the rocket payloads and a low-pressure version to keep moisture out of sensitive payloads. They tended to use several bottles of nitrogen in their attitude-control systems. The bottles held 7000 psi nitrogen, and through some pumping arrangements, they got that to 10,000 psi in the carbon-fiber balls inside of the payloads before launch.

When I arrived, we had dozens of bottles from three different vendors in Fairbanks. Taking over responsibility for them, I found we had no records of the serial numbers when we took them back. We were paying rental fees of a few dollars on each bottle and paying for the amount of gas that was used. I set up a system to log bottles in and out. Unfortunately, our records and the vendors' records left us with about a dozen bottles whose serial numbers we could not reconcile. For some we had to buy the bottle for $75 and keep it or donate it back to the vendor for free.

Liquid Nitrogen and Helium

In early 1972 we were preparing for the fall launch of a Black Brant V rocket and payload for the ICECAP mission. The Air Force Cambridge Research Laboratories scientists and payload engineers wanted us to be able to launch within 60 s by moving a launcher to the vertical position. The crew at Poker Flat made a plan for a new launchpad and quickly movable heated enclosure system for the 25-foot rocket.

This effort was a major element for the first Defense Nuclear Agency range improvement proposal in June 1972. Construction of what became pad 1 and pad 4 started a few weeks later with expedited funding from DNA. DNA had two MRL 7.5K launchers, consisting of a vertical pedestal with a cantilevered horizontal launch boom with rail, sent to Poker Flat from California.

The 60-second limit was predicated on servicing the infrared detector system with liquid nitrogen at the last possible moment, because the thermal vacuum bottle for the detector could keep the detector cold for only about 4 min, which was the flight time of the payload after launch. We met and exceeded the initial operational requirement for 60-s readiness. We built two structures to keep the rockets warm on the launchers before the launch season of 1972–1973.

The Space Dynamics Laboratory at Utah State University in Logan, Utah, built almost all of the infrared payload experiments for the ICECAP missions and follow-on DNA infrared programs that ran during my time at Poker Flat. They designed and had fabricated for them special liquid nitrogen and later liquid helium tanks called dewars that could keep infrared instruments in the rocket payloads cold for up to an hour.

Launch windows for experiments involving aurora remained several hours long each night as we waited for the proper science conditions. We designed and implemented heated rocket motor/payload enclosure systems for our launchers that allowed the Utah State crews to easily service their infrared payloads hourly while in the horizontal position.

In 2007 Ralph Haycock from Utah State reminisced about one particularly difficult infrared rocket-borne experiment. The temperature of the infrared sensor was approaching critically warm just as the aurora was getting better near the end of a long night. Ralph asked if we could hoist him and a helper and a canister of liquid nitrogen up in the basket of our cherry-picker truck to refill the dewar while the rocket was in the elevated position, instead of lowering the launcher and putting the arming system in safe mode, as was normal procedure before allowing anyone other than Carroll Coe near a ready-to-launch rocket.

A range safety rule is not to permit any electrical connections to be made or broken while the rocket is armed, but because there was no electrical switching involved in transferring liquid nitrogen to the payload, Carroll agreed to let our crew hoist the dewar so Ralph and his cohort could service the payload. Ralph told me that other ranges had muttered about bureaucratic safety concerns and had not allowed such events to take place, whereas Poker Flat was dedicated to science and more reasonable.

Payloads cooled with liquid nitrogen and liquid helium were challenging. When something was not working properly and we had to take an infrared payload off the rocket on the launcher, it could take up to a week before we were ready to attempt to launch it again. We had to let the liquid nitrogen/helium system warm up to room temperature, fix it, and then cool and test it before reloading it on the rocket on the launchpad.

I remember two instances when the mechanical structure that held the sensor platform together broke and the whole system had to be warmed up, taken apart, the mechanical part re-welded, and the system put back together and tested before we could remount the payload on the launcher. The first time this happened, the liquid-nitrogen-cooled mechanical mount had to be sent to Utah for repair, and we were down for about 4 days. The second time the mechanical frame block in a liquid-helium-cooled infrared interferometer broke. It was made of an exotic blend of metal structures with critical internal alignments. It would have had to be sent to Los Angeles for repair, which would have added at least 2 more weeks before we could attempt to launch it again. To everyone's amazement we discovered a welding shop in Fairbanks that specialized in the type of welding required, and we were ready to attempt a launch less than a week later after cooling and testing. When launched it worked perfectly, and we obtained great infrared data of the aurora.

It turned out that a similar mechanical structure was used in a piece of popular oil-well-drilling scientific instrumentation at Prudhoe Bay, Alaska. A welding shop in Fairbanks had learned how to re-weld and repair these structures. The Utah State folks could see the Fairbanks welding shop people knew what they were doing and were using equipment and techniques that the outfit in Los Angeles used.

US Army Meteorological Rocket Network at Poker Flat

In about 1959 the Atmospheric Sciences group of the US Army at White Sands Missile Range (WSMR), New Mexico, established and operated three meteorological rocket network sites to measure winds above balloon altitudes for the purpose of tracking atomic fallout. The northernmost site of the so-called Meteorological Rocket Network (MRN) was initially located at Fort Greely, Alaska. Their prime site was located at WSMR, and their southernmost site was located at Fort Sherman in the Panama Canal Zone. They established these as their contribution to an international cooperative effort to learn more about the structure of the Earth's atmosphere above 18 miles, the maximum altitude for standard balloon-borne measurements.

Each Monday, Wednesday, and Friday, these sites launched a standard Weather Bureau balloon to acquire wind speed, direction, and temperature data from the surface of the Earth to 18 miles altitude and then launched a small Arcas rocket. The All-Purpose Rocket for Collecting Atmospheric Soundings (a.k.a. Arcas) was developed by the Atlantic Research Corporation and deployed a radar-reflective parachute and temperature sensor connected to a small radio transmitter at 47 miles altitude. A critical measurement was always to make sure they gathered comparative data from both the ascending balloon payload and the descending rocket payload between 15 and 20 miles.

These data were sent immediately to the National Weather Service's Silver Spring, Maryland, facility, which was the US repository for data from 30 worldwide cooperative high-altitude weather measuring stations. Most high-altitude weather stations did not have a radar system to track the radar-reflective parachute. Their receiving antenna systems auto-tracked the radio telemetry signal, and the resulting azimuth-elevation angle versus pressure altitude data was analyzed to determine the winds aloft. The US Weather Bureau, now the National Weather Service, called this hardware a ground meteorological direction finder (GMD) system.

The army and other US high-altitude weather measurement programs got a boost when the air force became interested in high-altitude wind speeds and directions that might cause an intercontinental ballistic missile to miss its target.

Willis Webb at the University of Texas El Paso (UTEP) had the contract to do the science and some portion of the operations with the army's three meteorological rocket network stations. UTEP people were at Fort Greely but wanted the Geophysical Institute to take over their work.

In August of 1971, when I took over at Poker Flat, the Fort Greely MRN program was in the process of moving to Poker. The army team running the program would be stationed out of Fort Wainwright, just east of downtown Fairbanks. We had a contract to help the army build a new facility in support of their program but only for site preparation. The road leading through the range had been extended south to the base of the hill and up an 800-foot elevation change to the hilltop.

Neil Davis said that around June 30, 1971, with the end of the fiscal year and the upcoming Fourth of July long weekend, most GI faculty were gone. But Neil was still in the office, and as a result when GI director Keith Mather told Neil they had $30,000 he needed to spend right away, Neil had what you might today call a shovel-ready project idea. He wanted to use some of this money to get an electric power line from the substation on Poker Flat to the hilltop MRN site and the Chatanika ISR site at 27-mile Steese Highway. Neil called up a local contractor, who gave him an estimate of $15,000 to do the work. Neil was able to move the contract paperwork into quick reality, and the result was a three-phase power line that fed both the hilltop site and the Chatanika ISR from the Poker Flat substation.

John Miller, the guy I had talked to by ham radio from Thule before I came to Alaska, was the head of technical services at the Geophysical Institute. In 1971 he was responsible for one of the trucks we used to transport parts to build the radar dome from Moose Creek to Poker Flat, passing through a State of Alaska weighing and vehicle inspection station on the way. I was responsible for the other truck. John had a brand-new truck, and Poker Flat had an older truck sold or given to us by the Geophysical Institute. A ticket citing one of these trucks for lack of operating warning lights, license plate lights, or some such offense was placed in John's technical services mailbox. Shortly thereafter I received a memo from John about "Poker's" ticket, worded to indicate that I wasn't doing much to change his perception that the crew at Poker Flat was still a bunch of creative misfits who would steal or creatively borrow to build their precious rocket range. As you might have guessed, it turned out to be John's new truck that had earned the ticket. The truck we maintained passed the same inspection each day at the State of Alaska weigh station.

In midsummer, trucks arrived with custom-built insulated paneling for the 55-foot dome, which would be used as a control center for the program. The tie-downs on one load were undone, and that truck was driven up to the hilltop by a young army soldier. On a steep part of the road, the 20-foot-by-2-foot heavy steel panels with foam insulation in them started sliding off the truck one at a time and onto the road. The young army volunteer just blissfully drove on, not recognizing he was losing his load one panel at a time. Eldon was livid that the crew had to pick up each heavy panel, reload it, and haul them up the hill. He told the army team commander he never wanted to see that driver at Poker Flat again.

Notwithstanding this incident, Eldon was quite taken with these interlocking steel foam insulated panels. In the fall of 1972, we started designing a launcher shelter using the same kind of panels, which were custom-made for our purpose. They worked very well.

The improved MRN system would use a modified Nike-Hercules acquisition radar system with a co-aligned weather radio receiving system. Both the radar and receiving system could automatically track an airborne weather transmitter. I think the improved system at Poker Flat went operational in the winter of 1971–1972. In the command-and-control structure, all of the equipment looked first-class in modern clean facilities. In August of 1971, the US Army meteorological team established their boosted Arcas launcher and a Loki-Dart launcher on what is now pad 5, about 1,000 feet east of the blockhouse at Poker Flat. They also brought in a steel hut about 9 feet wide by 30 feet long and used it for storage and preparation of the MRN rockets and their payloads. They installed a standard GMD system nearby. The rockets were ignited by army systems installed in the Poker Flat blockhouse.

We continued to launch the MRN system rockets from pad 5 on lower range with ignition systems installed in the blockhouse. Our launch officer Carroll Coe always sat close to the MRN firing panels and helped the crew launch their rockets.

The US Army Meteorological Program was terminated in 1979 when satellite-borne systems became good enough to provide the data formerly available only from the met rockets. In a sense the satellite folks kept improving their systems using the MRN rocket data to calibrate their satellite observations. The day came when the satellites were doing this job so well 24/7 that the expense of maintaining the MRN people launching and analyzing data from relatively expensive balloons and rockets was no longer justifiable.

Dynamite Duo Takes Down the A-Frame

The legendary A-frame, the large protective weather shelter on pad 1 built for the first rocket launch in March of 1969, sat unusable on frost-heaved rails in August of 1971 when I came to Poker Flat. The heaving permafrost beneath the rails the launcher rode on had made it difficult to move the A-frame after its first year of use. Eldon, Neil, and Larry were in the final stages of building lightweight clamshell-like covers attached to the launchers themselves.

In April of 1972, we proposed to build a Payload Assembly Building. We assumed we would recover and reuse about 90% of the materials in the A-frame for this building. Eldon presented me with a plan to drape dynamite on detonating cord from the apex of the A-frame down the walls inside and across the semiopen canvas front. He said this would pop all the metal sheeting and plywood off and bring all the metal and wood framing to the ground, so we would recover it for reuse.

Eldon and Jim Woolf, our wind-weighter, had State of Alaska blaster's licenses specifically for dynamite. Eldon had gotten his for some work around the house years before, and Jim had a military blaster's license. To get a blaster's license, they said, you had to make an appointment with the appropriate people in the Division of Mining for the State of Alaska and tell them what you knew and answer any questions they had. For example, they wanted to make sure you knew how to use an electrical cap or a fuse to set off a stick of dynamite and how you would do this with a half-stick or multiple sticks held together with bailing wire. Eldon and Jim had held licenses for several years, and apparently there were also no follow-up conversations through the years as improvements were made with how to set off dynamite.

No one else on the range crew wanted anything to do with this effort, which should have been a major clue to me that perhaps Eldon and Jim did not really know what they were getting into. But I told them to go ahead. I told Eldon that he and Jim would be the only ones involved, and they were fine with that.

Eldon and Jim placed the charges. Using long runs of Primacord, they made a simple wrap around half and full sticks of dynamite, which they hung from the interior of the A-frame. They placed explosives in a similar manner along the rear solid wall. The front of the A-frame had been enclosed by a large canvas wall with the world's biggest zipper (or so it was claimed) running more than 70 feet from the A-frame apex to the ground. Eldon and Jim rigged explosives into this relatively open area and declared they would set them off first to create a blast wave that would contain the blast when the other explosives went off, thereby causing the whole structure to collapse. Eldon and Jim spent days carefully placing these explosives.

Aerial view of Poker Flat taken in the summer of 1974. The A-Frame launcher cover is shown in the lower left, over old pad one, before it was destroyed

I hired Al McNeil from the Geophysical Institute photo group to come out and take pictures and movies of the explosive destruction of the A-frame.

The countdown went smoothly, and the A-frame puffed out just as Eldon had said it would. It was a spectacular event. Unfortunately, it did not create the expected nicely organized pile of rubble we could easily pick through for reusable materials. When we walked over for a closer look, we saw several sticks of unexploded dynamite with Primacord wrapped around them in the chaotic pile of lumber. I got that "we told you Eldon and Jim did not know what they were doing" look from almost everyone else on the range crew. No one was willing to go into the rubble until Jim and Eldon found and removed every bit of unexploded dynamite and Primacord, and so that is what they did for the next several days. Neither expressed to me that they felt this recovery of unexploded dynamite was dangerous.

I logged the destruction of the A-frame in the official launch list of Poker Flat as "PF-af-39A, GI/A-Frame Destruction/Brown, UAF," occurring at 0235:00 Coordinated Universal Time on April 7, 1972.

Unloading Rockets from Fort Churchill

One winter, NASA decided to fly three Black Brant rockets from their storage at Fort Churchill, Canada, directly to Fairbanks for almost immediate launching at Poker Flat. We knew when the air shipment was coming and made arrangements with the FAA and the Fairbanks International Airport for the plane to land and park in an "explosives safe" area.

I put Merritt Helfferich in charge of arranging for the rockets to be offloaded and brought to Poker Flat. I went to the airport with Merritt to meet the plane and crew. The plane was an old Constellation with tripod landing gear, one set of wheels under the nose and another set under each wing. The body of the aircraft was at least 20 feet off the ground. The cargo doors were in front of the wing, behind the cockpit and to the rear of the wing on one side. I could not imagine how the 20-foot-long wooden crates containing rocket motors could be onboard, but they were.

Merritt and I paid a visit to a Weaver Brothers representative who loaded and offloaded freight from planes and invited him to come look at what we saw as a huge problem. Did I mention that it was minus 20° outside? He laughed and said he loaded and offloaded things like this all the time. We hired him on the spot and waited at the airplane to watch.

He returned with the biggest forklift I had ever seen—I swear it was about 30 feet tall—and a 20-foot-long piece of thick-walled pipe large enough in diameter to fit over one of the forks. He laid the pipe on the ground and drove around so that he could run a fork inside of the pipe. He maneuvered the forklift in front of the wing; lifted the long pipe, which he called a "donkey dick," into the air; and oh-so-carefully inserted it through the cargo door at a very shallow angle. He got an amazing amount of the pipe inside the airplane and lowered it close to the top of one of the rocket boxes. The flight crew removed the tie-downs that held the rocket box to the floor of the aircraft and strapped the rocket box to the long horizontal pipe. The forklift operator lifted the rocket box a few inches as if it weighed nothing and started backing away from the aircraft. It took the operator a couple of maneuvers back and forth until he had the rocket box firmly tied to the pipe the way he wanted it before pulling it entirely out. He put the rocket box onto the enclosed and heated truck van we had hired to haul the rockets from the airport to Poker Flat. I think he offloaded the airplane in less than 30 min.

Later, the management of the Fairbanks International Airport called Merritt to complain that several of their blue warning marker lights surrounding the area where we had offloaded the airplane had been destroyed. We gladly paid for the damages.

Storing Rockets

Parts and Assembly

For the first launches in the spring of 1969, rocket motors had been stored, uncrated, and assembled at Fort Wainwright, then transported to Poker Flat fully assembled for installation on the launcher. The barium payloads for the first launches were assembled in Rocket Assembly Building A, known as A-RAB, at Poker Flat. In

1970 and 1971, rocket motors in their packing crates were stored in a Jamesway tent that sat right on the ground.

In the spring of 1972, I set the range crew to build Rocket Assembly Building B (B-RAB). It would be of modern wood-frame construction, with an insulated concrete floor and a large overhead door, and be heated by an electric hot-water baseboard system. It was the highest priority for the crew at Poker Flat.

Soon after B-RAB was completed in the summer of 1972, the Defense Nuclear Agency ICECAP program came into being, and in 1973 we were funded to get an architectural design for a new rocket storage building. The storage building would be built where the range road started up to the Meteorological Rocket Network site at the top of the hill. We wanted a hydraulic platform inside the building behind the overhead door where trucks carrying rockets could back up to the building.

A bulldozer clearing the site ran into hard schist rock where we wanted to place the hydraulic system for the platform. Eldon and Jim came forward again, telling me this was a small dynamite project that they could successfully do. Despite having a bad feeling about their prowess in doing things with dynamite, I authorized them to break the tough rock with dynamite for the hydraulic platform. This time they drilled a number of vertical holes in a roughly 6.5-foot-square site and loaded dynamite into the holes to a depth of about 10 feet. After the dynamite was set off, we gathered around and peered into the rubble and saw that there was considerable unexploded dynamite in the hole. Eldon and Jim kept their word and carefully extracted it all by hand. This was very slow-going and took about a week.

The architect had designed a bunker with concrete walls and roof. I don't remember if we had a contractor involved to pour the concrete slab and walls, but by the time we got the walls up about 6 feet, we knew we would not have enough money to complete it as a concrete structure. We built the rest of the way up with wood framing and a wood-frame truss roof.

We stored ignitors and explosive ordinance in a small wheel-less trailer beside B-RAB from 1972 through 1989. The only giveaway that anything worthwhile was stored in the trailer was the lock on the door. Once inside you would see open wooden cabinets along one wall like you might see for plumbing parts and scattered around on the floor and shelves were cardboard boxes partially filled with carefully wrapped items. Booster motor ignitors typically came in sealable metal canisters and were stacked neatly here and there. There were a few small rockets a few inches in diameter by perhaps as much as a foot long on at least one shelf. There were spin motors that were placed around the headcap of an Honest John rocket.

A sounding rocket payload may use several interesting explosive-actuated devices. Despin weights are wrapped around the outside of a payload section as a pair secured by a cable through the center of the payload. An explosive-driven piston with a knife edge on one end is used to cut this cable. Specially designed clamping rings hold payload sections together. A similar explosive-driven knife edge is used to cut the bolts that hold it together. The clamps are spring-loaded to fly away, separating one payload section from another. Similar explosive-driven knives cut small bolts that secure spring-loaded doors so they deploy at altitude. Small rocket motors are used to pull the nose cone away from the rest of the payload at altitude. Each rocket that came to Poker Flat was accompanied by a separately shipped box of the necessary explosive ordinance for ignition of the rocket motors and actuation of unique payload devices.

Keeping Assembled Rockets Warm

To keep assembled rockets and their payloads warm, the Poker Flat crew used portable heaters that blew heated air through a disposable hose into a bag made of Velostat, a thick, tough, black plastic sheeting. The heated air kept the bag inflated like a long cylinder balloon until the rocket tore it to pieces during launch.

A couple of years into my tenure, Beau Battey brought me a chunk of Velostat that consisted of several layers of semimelted plastic formed into a V-shape about 8 inches long and an inch on a side. He told me he had removed it from the leading edge of one fin of a Nike rocket that had impacted across the Chatanika River.

We had long worried that the Velostat might catch on the fins during launch. Nikes take off quickly and seemed to rip the Velostat bag to shreds. But it happened so quickly that we could not see it. We worried that the roar of flame out the rear of the rocket would suck the plastic in around the sides as it thrust upward, and here was proof that it sometimes did. The Velostat might have been causing problems with regard to fin settings and spin-up of the Nikes. Sounding rocket engineers go to great lengths to measure and set fin angles on their rockets prior to launch. There was no question that a slab of plastic would spoil the aerodynamic characteristics of a fin. The bigger question was if it could cause the rocket to fly off target after launch so that it landed somewhere we did not want it to. We had not seen any obvious examples of such a deviation but decided to do everything we could to keep the plastic from gathering on the fins of future launches.

On launchpad 2, with its Sandia Corporation HAD launcher, the range crew had crafted a lightweight metal box hinged along either side of and along the launch boom. The launcher formed the top of the box, and L-shaped pieces hung down that formed the side and half of the bottom of the box. The box held commercial sheets of 1-inch-thick Styrofoam, and it was covered with a wrapper of clear Visqueen. I think heat was provided by a gasoline-fired heater on wheels. It was a huge volume, but it worked fairly well. The Visqueen kept winds from blowing the Styrofoam apart and kept the heat in. An electric motor and gears could rotate the two halves of the box up and away from the launcher or close the box around the rocket on the launcher.

In 1972, Jim Heppner brought a rocket-borne chemical release program to Poker Flat that required more launchers than we had. NASA moved its Rocket Aft Grab (RAG) launcher from Barrow to Poker Flat to support Jim's effort. It literally held the Nike only by a small grip on its nozzle. There was about 2 inches of guided travel after ignition of the rocket. It was designed to launch only Nike-boosted, relatively lightweight upper rocket stages and payloads.

The RAG launcher arrived with two installation possibilities: pour a concrete base with a ring of mounting bolts to match the base of the circular tub of the launcher, or set it up with its mounting legs on a flat space. The RAG had a frame that held three long beams with leveling screws at the end. Two beams were to be set up so they encompassed the possible launch azimuths, while the third was installed to the rear. We had no time or money to do anything but use the leg approach to mount it. We mounted it on the meteorological team's concrete pad. With his typical "I can do that" approach, Eldon poured successive layers of water mixed with sawdust over the legs, freezing them into place within a day or so. Each launch would blow some of the ice/sawdust away, but it was easy to add new material and be ready to go within a day.

We built a real kluge of an overhang with canvas sides behind a trailer and backed it up over the RAG launcher. The rockets and payloads were enveloped in Velostat bags, and everything inside was kept warm by a gasoline-powered hot air heater.

With ICECAP funding we installed two new MRL 7.5K launchers. The original pad 1 now became pad 2. A new pad 1 was erected to the west of the original pad 1, and a near identical pad was installed to the east, which became pad 4.

We designed and built more elaborate systems to shelter the rockets and payloads on the newer launchpads and on our original pad 1. Our approach to keeping rockets and payloads safely warm kept evolving through the years. Ultimately we bought our own Styrofoam-spraying machine and 4-by-8-foot sheets of white Styrofoam and created 8-foot-long U-shaped boxes that clamped into structures we permanently welded along the booms of each launcher. We used shish kebab sticks to hold Velostat and Styrofoam access panels in place in these new systems. I can't say for sure that we invented this idea ourselves. It may have had many fathers, but it seemed to work well.

Creating a Clean Room

We bought a small clean room to enable technicians to assemble sensitive telemetry and other electronic systems in rocket payloads without fear of dust ruining the circuits. The contract called for an installer to help set it up. I learned that because the airflow was from ceiling to floor, only the very upper surface of the payload would be clean, and everything below it would be "dirty." It was a well-lit clean room with no measurement devices in it. It was a class 10,000 clean room, which means less than 10,000 particles 5 microns in diameter per cubic foot of air space. Although better than what we had before, range users and scientists wanted a class 100 clean room, which maintains less than 100 particles larger than 0.5 microns per cubic foot.

US Air Force Cambridge Research Laboratories had a much bigger clean room at Churchill, which they reluctantly shipped to Poker Flat. We took down the small clean room and put in the much larger clean room from Churchill. This time we were asked to drape static-free black Visqueen around the sides. But there was still no measuring device. The air was down-flow. This room was also rated class 10,000.

Soon we took it down, turned it crosswise in the Payload Assembly Building, and enclosed it with quarter-inch plywood clad in metal for easy cleaning. We built it such that cleaned air circulated down, then back up the sides, and over the top to be pushed down again. Utah State University showed up with an ultraviolet light and camel-hair brushes. They shut off all the lights in the clean room and found the dust with UV light and brushed it aside with the brushes.

Later, after I left in 1989, Poker Flat added a whole new building on the northwest side of the Payload Assembly Building that included a hallway around the outside, an entry cubicle with sticky floor, and small boxes through the walls so that you could put a requested tool in from the outside. It was arranged so there was no exchange of air. This clean room was professionally built and installed. A very sophisticated measuring device inside looked like an oscilloscope with a long, small hose. This new clean room also had humidity control, something we never accomplished in the previous clean rooms. Our dry climate in winter left dirt particles on the payload surfaces because of static electricity. This class 100 clean room is still in existence today and well used.

The Evolution of Radar Systems at Poker Flat

Radio detection and ranging (radar) involves the use of a radio signal turned on and off rapidly enough that the time-of-flight of the signal reflected off a metallic target like an airplane can be accurately measured in milliseconds to microseconds. (Milli- is one-thousandth of a second and micro- is one-millionth of a second.) In the case of the ionosphere, there is no hard metal target, but there are relatively compact electron density layers at different altitudes. These are caused by ultraviolet sunlight tearing electrons loose from atoms and molecules during the day. It takes a long time at night for the electrons to recombine with an atom or molecule. The time for this recombination rate depends on altitude. The lowest layer is called the D layer, at about 40 miles. The E layer is at about 75 miles, and the F layer is from 93 to 300 miles. To measure the altitudes, the ionosonde transmitter energy is directed into a vertical antenna array.

In the summer of 1965, I had slept in the ionospheric sounder, or ionosonde, room that was yards away from the Geophysical Institute in the Chapman Building. It ran 24/7 and every 15 min put out a high-power burst of sequential frequencies from 40 to 150 MHz that made a huge noise on my AM radio. Although I was focused on my observational studies of the near-infrared emissions of the aurora, I learned about the ionosonde and how it could be used to detect overhead auroras.

Also, with a dish or antenna system aimed toward the northern horizon, it could detect intercontinental ballistic missiles.

My next encounter with the principles of radar involved sound reflected from temperature inversions during the winter in Fairbanks. These temperature inversions created boundary layers that reflected sound waves. A sonic detection and ranging (sodar) transmitter pulsed at about 2000 Hz about ten times per second in a square box with the loudspeaker aimed upward. A separate loudspeaker colocated next to the transmitting speaker was the receiver. The temperature inversion layer was typically somewhere around 300–900 feet above the ground. I got so interested in this project that I wrote a proposal for students at West Valley High School in Fairbanks to make one after I toured a sodar facility operated by the National Oceanic and Atmospheric Administration north of Boulder, Colorado. NOAA was measuring sound echoes from air density unconformities created by winds coming from west to east over the Front Range mountains. West Valley High School put up a 30-foot-tall tower with a temperature, wind speed, and wind direction system on it. The project went well but lasted only a year and a half.

In 1972, Poker Flat relied on the FAA radars on Murphy Dome west of Fairbanks to tell us if we could launch a sounding rocket. Poker Flat is in the radar shadow of Pedro Dome. Poker Flat had a 2200 MHz NASA VERLORT radar system that was mounted in a 40-foot trailer. It was optimized to work with a radar beacon on a sounding rocket payload. Upon hearing the VERLORT radar, the beacon would wait 5 ms before blasting a 300-watt signal at the same frequency back to the VERLORT radar. The timing, in the microsecond range, between the VERLORT transmitter pulse and the pulse received back was used to detect range. In this configuration you could track a hard metal target like a first-stage rocket booster out to about 12 miles.

In the spring of 1972, we had so many launches lined up that NASA brought in a trailer-mounted MPS-19 radar. It stayed with us for 2 or more years.

The grounding system at Poker Flat kept us from running a data cable from these radars to the blockhouse 1000 feet away. A lot of noise, masking the real signal, kept appearing on the data cable. In 1972 Bob Spies and I went to White Sands Missile Range, where Bob saw a late World War II antiaircraft radar called a T-9. WSMR had a junkyard full of them, so they gave us one. We set it up near the blockhouse for wind-weighting data to six miles. It used synchros connected to synchros in the antiaircraft gun so they followed one another within 1/100 of a degree in azimuth and elevation. Bob figured out how to build an isolation system so we did not have a grounding problem connecting it to the wind-weighting computer in the blockhouse. He built another isolation system and connected it between the VERLORT and the MPS-19 and the wind-weighting computer in the blockhouse.

Poker Flat Temperature Inversions

In December 1971, we had only two wind direction and speed measuring units on our 265-foot-tall meteorological tower at Poker Flat. We borrowed two more, so we had four on the tower in time for the first launches in February 1972. The units were of the World War II variety made out of heavy metal. The propeller-generated voltage indicated wind speed and could be read out on a meter. The direction came via a synchro that drove a dial in the blockhouse. It took a lot of work to read these and enter data for wind-weighting rockets. Each summer we had to replace the bearings in both the wind speed and wind direction units.

With our range improvement proposal, we got money to get six new digital wind sets for wind speed and direction. These worked great, and we interfaced them as digital inputs to our Interdata Model 4 minicomputer. We had to take them down before lightning season, early June to the end of August, because even nearby lightning strikes had in the past destroyed the wind sets.

I was interested in sound detection and ranging of temperature inversions in Fairbanks, and I had our people install thermistors next to each of the wind sets. I noticed that if we had a wind coming down the valley, temperatures on top of the tower started warming up. This progressed over a few hours until it was also warm on the ground. The wind direction also turned up the valley when temperatures warmed, and at each level of the tower as the temperatures warmed, the wind direction turned to measure winds coming from up valley.

We kept an A-frame ladder near pad 1. One day, Eldon Thompson was doing some work on a launcher and said it was so warm he threw his coat to the ground. Where I stood, it was −10° and there was no wind. He said there was a gentle wind up on the launcher. We swapped places. I climbed the 16-foot ladder and felt the temperature rise by 40°.

Of course, I had to investigate. I learned that when a high-pressure area built up over the Yukon Flats, temperatures plummeted in the Fort Yukon area, like they did at Poker, to 40 or more below. When the height of the temperature inversion line equaled the elevation of Eagle Summit 60 miles north of us, air spilled downslope to Poker, an elevation change of about 1500 feet. In doing so, the air warmed up due to pressure expansion.

We were familiar with winds out of the north causing our rockets to dive into them during their powered portions of flight. I learned that the effect started around 50,000 feet, moved to the ground within a few hours, and persisted for up to a week. These winds were a wind weighter's worst enemy, because they frequently caused many a long night with great auroras, clear skies, and no launch.

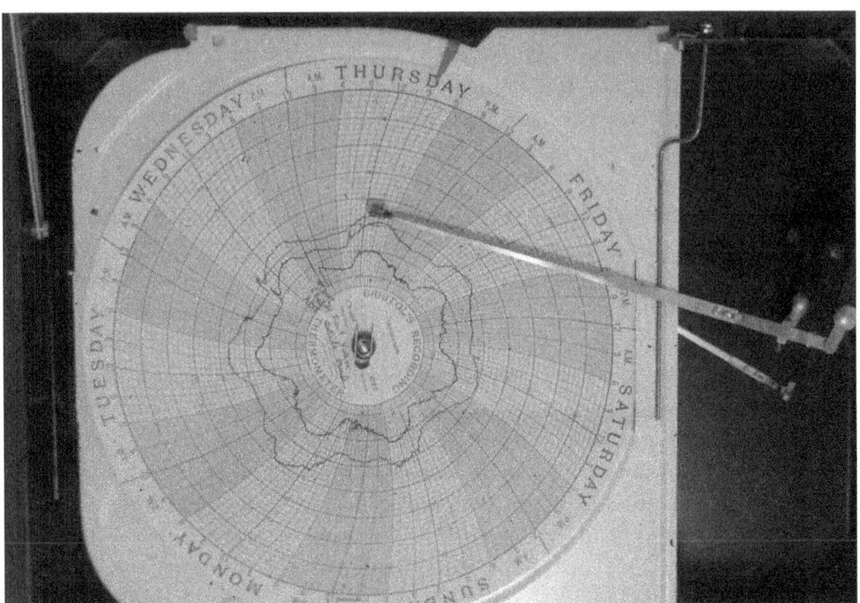

During the 1970s, Poker Flat personnel relied on weather readings from Fort Yukon, Alaska, 145 miles northeast of Fairbanks to predict weather conditions ahead of planned sounding rocket launches

NASA took the MPS-19 back to Wallops Flight Facility in 1974 after upgrading the VERLORT with a data acquisition computer. We used NASA's VERLORT systems until about 1985, when NASA informed us they had put out a bid document for a C-band radar that used the Nike-Hercules radar pedestal. The digital range radar used in all the air force and NASA installations cost $18 million to make and $200,000 a year to maintain, requiring a 3-year contract. NASA Wallops informed us they were going to make five of their own and that they would maintain them. The digital range radar arrived at Poker Flat in 1985. Electronically it was fine, but the metal structures used to hold the dish, horn, etc. were broken from the transport across the United States and over the Alaska-Canada Highway.

Wallops asked us to build a platform strong enough to take the stress of azimuth and elevation motion without deviations while tracking a sounding rocket. We built the platform using three columns of surplus 48-inch pipe from the Trans-Alaska Pipeline. We got praise that it was more stable than they had at Wallops, at Andoya in Norway, or at White Sands Missile Range. It also had a built-in optical tracker.

After I left Poker Flat, NOAA put a coherent Doppler radar on top of Pedro Dome that could detect wind shears and Doppler-shifted data indicating rain, wind speeds, and directions over the top of Poker Flat. It was a multimillion-dollar project that is still in use. Today, with high-speed Internet, all wind-weighting data at Poker Flat is sent to Wallops Island, Virginia, to calculate launcher settings at Poker.

The next installation was the mesospheric-stratospheric-tropospheric (MST) radar, which constantly monitors wind speed and direction at altitudes from 6 to 60 miles via a 325-square-foot antenna field operating at 50 MHz and 15,000 watts. The radar system operation, spearheaded by Ben Balsley of NOAA and operated with the help of Poker Flat personnel, is an innovation in middle atmospheric research. Poker Flat had an abandoned shallow gravel pit on the land it got from the Fairbanks North Star Borough and got a contract to fill in the pond left behind. NOAA installed 50 small transmitter huts, each with 15,000-watt transmitters operating in a coherent fashion at 50 MHz. A building off to one side housed the computerized phasing for the cables to the huts, which was delivered to a small computer. The MST system operated from about 1975 through 1978 and detected a stratospheric warming, representing an innovation in middle atmosphere research.

Building a Science Observatory

In early December of 1975, I got a call from our Defense Nuclear Agency contract monitor Herb Mitchell, asking if, for $65,000, we could build a permanent science observatory at Poker Flat in time for the February ICECAP rocket launches. This observatory would be built to house riometers, magnetometers, and other instruments used in routine Geophysical Institute experiments, plus all-sky cameras and meridian-scanning photometers used to support rocket launches.

Herb knew all about the prelaunch science information we acquired from our Ester Dome and Fort Yukon Observatories and communicated to the Poker Flat blockhouse to help rocket scientists make good decisions to get their rockets successfully launched. For the past 2 years Herb had funded the transportation of an Air Force Geophysics Laboratory (AFGL) trailer to Poker Flat and back to storage stateside, which he now told me had cost $45,000 each time. AFGL had modified this trailer to become a small aurora observatory and command post for their ICECAP launches. Herb told me that the trailer had originally been built to support the 1962 atmospheric nuclear weapon tests from Johnson Island in the Pacific, and it was incredibly heavy because it was lined on all sides with lead sheeting. AFGL was again asking for money to bring it up for their ICECAP launches in early 1976.

Herb and I had talked in the past about how the growing popularity of Ester Dome as a place to install high-power radio station antennas was creating light pollution and external electronic noise that would eventually overpower our ability to continue using Ester Dome as one of the main GI optical observatories. I tried to talk Herb into letting us build the observatory after the spring 1976 launches. I pointed out the practical difficulties associated with building such a facility so quickly in the winter and that we would get more bang for our buck, build a better observatory, and have something to do in the latter part of our fiscal year. But the only way he could avoid funding AFGL to bring their trailer to Poker Flat for the February launch program was to give us their money to get an observatory built and working.

Without giving much more thought to it, I told Herb we could get an observatory built on top of the hill near the MRN facility for $65,000. Herb added that money to our range-operating contract in time to cover the expenses. AFGL might have expressed their subsequent dismay that Herb had diverted funds for moving their trailer to build a permanent science observatory at Poker Flat, but no one from AFGL ever called me or asked what we were going to build.

When I informed the Poker Flat crew that I had told Herb that we could clear a site at the top of the hill near the MRN facility and have a permanent science observatory ready to use for the February launch program, they immediately sat down with me, and in less than an hour, we figured out how to do it. Eldon Thompson was of course instantly enamored of and led all aspects of this project.

I put Dan Osborne in charge of the science needs of the observatory and asked him to take the lead in building and operating the observatory. Dan made sure Eldon knew and made arrangements for the roof openings we would need to install launch support scientific instrumentation. Later, we only had to cut openings in the prelocated and framed holes in the roof.

Eldon started in right away. He hired a big bulldozer and leveled a nice dark place near the top of the hill behind Poker Flat. We ran into some tough rock and elected to dig holes and use pilings to support our building, which was prefabricated in sections in the Payload Assembly Building. We had the site cleared and wood pilings in place before Christmas 1975. By early January 1976, we had electrical power at the site and were hauling modular insulated wood-frame floor pieces up the hill and placing them on top of the wood pilings. By mid-January we had a prefabricated, insulated 30-by-60-foot floor on top of the pilings, electrical power, and communication. We had 8-foot walls and a flat roof installed by the first of February. We built holes into the roof sections for future domes but covered them over. Some of the equipment in the observatory was moved from the Ester Dome field site, which for economy's sake was closed.

We would need to lay a communication cable between the communications shack on the lower range next to the blockhouse to the new observatory, and this took an unexpected turn. The cable was on a reel on a trailer near the future observatory, and we tied a rope to one end and the other end to our Snow Trac vehicle. The Sno Trac slowly pulled the cable off the reel and behind it down the hill. A few hundred yards downhill, this cable would cross the road leading up the hill. I got a call on the radio in my office to come up and bring my camera. When I arrived where the cable was to cross the road, the Sno Trac sat on the road and the cable it had pulled lay in a tangled pile just uphill of the road. We had not thought to secure the upper end of the cable. Although it had been moving slowly, it had a certain momentum in the icy groove it made in the snow, and it kept on coming down the hill after the Sno Trac had stopped. I got our electronic technician Joel Lindsey to crawl underneath and in the loops of cable and took his picture. The cable itself was about 2 inches in diameter and had a few hundred pairs of communication wires inside of it. We used the Sno Trac to pull the whole cable back up the hill to the observatory, tied the upper end off so it could not slide down the hill, and did it again.

Jack Winckler from the University of Minnesota was the first to use the new science observatory. His launch involved Pamela Roth from England and two of her graduate students, who helped Dan cut panels out of the roof to expose the framed areas Eldon's crew had built for us to install observing domes. Dan got all the equipment from Ester Dome installed and operating in time for the launches, and Jack's Black Brant flew on January 31, 1976.

We needed some way to stay warm and see the sky. The building was built so that the long axis was magnetically aligned east to west. Along the north side of the building, we had the range crew install a waist-high 6-by-4-foot shelter with its own electric heater. It had a roof that slanted roughly 45° up the side of the building to attach near the roof. On the slanted area, we cut out a round hole and installed a 5-foot-diameter Plexiglas dome. One could now step through a door in the side of the observatory and stand in a completely dark area and observe the aurora with your naked eye. The British graduate students immediately told us we should call this the Loo, for the place you find a toilet in England. The name was too good to pass up. One day my daughter Melody came to Poker Flat with me and put the first coat of paint on the inside of our Loo.

You could stand in this room and look at the aurora magnetically east, west, north, and back overhead. At most, two people could be in it and comfortable at the same time. We equipped it with heat and communication, and it served as the last-minute eye on the sky launch decision center.

I don't remember when we hired a contractor to drill a well at the science observatory, but it was early in its history. Geologically, we knew the observatory was built on top of Birch Creek schist. The driller struck a dribble of water at 180 feet and gave me a Styrofoam cup of Birch Creek schist grindings from that level. He continued drilling down 450 feet through Birch Creek schist until he ran into what he described as a loose layer. He was drilling through relatively hard material and blowing his grindings back up to the surface with air. Once he got into the loose material, he lost his air into the loose material and could no longer bring up his grindings.

The option would be to case the well with metal pipe before he could drill deeper. This would be expensive, and to his mind not worth it. We paid him off, and he packed up his drilling equipment and left. We opted instead to install a thousand-gallon water tank and have water commercially hauled from Fairbanks to Poker Flat, and we installed a composting toilet.

At some point I discovered an unused astronomy-type metal dome that had been installed at Gilmore Creek to support the Baker-Nunn cameras to track satellites. The Baker-Nunns did not become operational until well after the Minitrack radio tracking systems, which could operate in cloudy as well as clear weather, took over providing tracking data for satellites. That was followed by dish tracking systems. The Baker-Nunn had been sent to the Gilmore Creek satellite dish tracking site. But Ed Eisle, NASA manager for the site, told me that it was filled with cosmoline grease and that they never got it degreased enough to work in cold weather. The dome had been surplused at Gilmore. I discovered it being used as a shelter in a dog yard near North Pole, Alaska. I bought it for about $600 and had it installed on top of a structure Eldon had built on the science observatory.

By 1976, the Geophysical Observatory at the top of the hill included a magnetometer, a riometer, all-sky photographic cameras, an all-sky television camera, a meridian-scanning photometer, and an elaborate optical spectrophotometer system. Readouts of instruments as far away as Fort Yukon, 125 miles to the north, were available to give information about the geophysical environment downrange, and some readouts were made available to the nearby Chatanika radar site. Each year, as time and funds permitted, we made improvements to the facilities. After interior and exterior finish work on the Geophysical Observatory building was done, including installation of sewage facilities, we constructed a permanent shelter for the large optical tracker system adjacent to the observatory.

The Geophysical Observatory was seeing increasing use by local and visiting groups, including several from Japan and the United Kingdom. It was also a key site in the Alaska meridian chain network established as part of the International Magnetospheric Study.

The Geophysical Institute acquired large amounts of ground-based data on the geomagnetic conditions before, during, and after launch for most rockets from Poker Flat. In particular, we produced summaries of data relevant to the ICECAP rocket program, intended to study changes in the upper atmosphere associated with geomagnetic activity. These reports summarized the general meteorological conditions; the global and local magnetic activity; all-sky camera observations of auroral morphology from various stations in Alaska; photometric information on the intensity of the aurora from Fort Yukon and Ester Dome or Poker Flat; riometer absorption at Fort Yukon, Poker Flat, and College, Alaska; and multifrequency ionosonde data from College. In addition, the data were interpreted to provide information on the phase of the magnetospheric and auroral substorm prevailing during the rocket launch.

In 1987, Geophysical Institute Director Syun Akasofu and I approached Alaska Senator Ted Stevens to get funding to upgrade Poker Flat as a "national facility." Syun referred to the observatory at Poker Flat as a wood-frame rabbit warren that might easily burn down. The upgrade money took years to materialize and came after I left in 1989. Syun led the effort for the modern, well-equipped observatory that now exists at Poker Flat and had it dedicated as the Neil Davis Observatory.

Chapter 6
Memorable Events, Early 1970s

For a typical midwinter nighttime attempt to launch a rocket into a particular kind of aurora, the first crews came on station about 5 p.m. A typical nighttime launch window goes from 7 p.m. until 3 a.m. If launch criteria of aurora, weather, or payload readiness were not met, we scrubbed the mission around 3 a.m., and everyone went home or to their motel rooms in Fairbanks. We often counted down and scrubbed night after night.

Twenty-two rockets were launched from Poker Flat in 1972. In addition to the Infrared Chemistry Experiment Coordinated Aurora Program (ICECAP) mission, Rice University returned with two payloads. Jim Heppner of Goddard Space Flight Center flew 3 more of the 30-some missions, most of them chemical release explorations of electric fields, which he would fly over the next 8 years.

The Geophysical Institute and Los Alamos National Laboratory (LANL) undertook three more barium releases, all coded with the innocuous names of South American birds. This series was particularly challenging as it required coordinated observations from Alaska and New Zealand.

In conjunction with the solar eclipse that summer, which was visible in its totality north of Poker Flat, LANL, and Sandia launched a payload designed to measure ultraviolet solar emissions, while a group from Atmospheric Sciences Laboratory measured ozone levels in the upper atmosphere and compared them with normal levels.

The Geophysical Corporation of America performed some pioneering lithium release experiments during which thermospheric wind fields were measured in daylight hours. Other experiments included the Universities of Minnesota and New Hampshire's measurements of electric fields aligned along the Earth's magnetic field lines.

N. Brown, *Northern Lights and Rocket Flights*, Springer Biographies,
https://doi.org/10.1007/978-3-032-14598-7_6

The Missing Igniter

Sometime in the 1970s, near the end of the 2-week-long prelaunch preparations of a Goddard-supported sounding rocket at Poker Flat, I was told that the initiator for the igniter for the first-stage booster rocket was missing. Further investigation revealed that it had never been shipped. The moon-down window—the 2 weeks that would be optimal for launching the rocket into the aurora—was starting, and the principal investigator was really upset.

The initiator part of the igniter assembly that was needed was no more than a couple of inches in diameter and a half-inch thick, but it was rated as a class C explosive. Inside this assembly were the equivalent of two model rocket igniters: two small wires coated with chemicals that burst into flame when the wires were electrically heated by the firing current. These were miniature flame producers, not explosives. When fired, the flames might occupy at most about one cubic inch of space and would not produce explosive gas products. This initiator screwed into an igniter assembly. When the initiator wires burned, the main charge ignited and flushed flame through the guts of the sounding rocket motor, setting it on its way.

An initiator was appropriately wrapped, surrounded by considerable packing, enclosed in a metal can, and put into a 12-inch-square wooden box, appropriately labeled. This was sent to a commercial air carrier for transport from Goddard Space Flight Center in Maryland. By now the package weighed over 10 pounds and was ominously labeled "Explosives," which got the attention of anyone handling it along the way.

Class C explosives can be carried on passenger flights at the pilot's discretion, and it took a day or 2 before a pilot agreed to bring it onboard a 747 aircraft flying from Baltimore, Maryland, to Anchorage, Alaska. We made arrangements for another Alaska Airlines flight to bring the package to Fairbanks.

Tom George would accept the package at the Fairbanks airport on our behalf and bring it to Poker Flat as soon as it arrived. Tom, who was working at the Geophysical Institute at the time, had worked for me at Poker Flat, hauling construction supplies the 30 miles from Fairbanks to Poker. He served as an expediter: we would give him purchase orders, and he would take them to the Geophysical Institute, get them signed, fetch the supplies, and haul them to Poker Flat each day.

As I recall, we loaded the first- and second-stage rocket motors and payload onto a launcher at Poker Flat and checked everything out. All we needed to launch the rocket was aurora and the initiator installed in the first-stage booster igniter.

The flight from Baltimore to Anchorage was scheduled to arrive in the early evening. We were notified as soon as the plane took off from Baltimore, but a couple of hours before the flight was due to land in Anchorage, we received a phone call telling us the plane had landed in Fort St. John, British Columbia. A passenger had had a heart attack, so the plane made an emergency landing at one of the few runways along the flight path long enough for a 747 to land.

Our plan to have the package transferred from the Anchorage inbound airline to one heading for Fairbanks that same night fell apart, because the transfer office would be closed by the time the package arrived. So we bought Tom a round-trip ticket to Anchorage and arranged with the inbound airline for Tom to be authorized to pick up the package. Tom's challenge was to talk the pilot of the Fairbanks-bound plane into allowing the package onto the passenger-carrying flight.

As expected, the explosives labeling on the package raised a red flag with the airlines in Anchorage. Fortunately Tom, himself a pilot, was able to speak directly to the captain. The pilot was concerned that they would be flying right by the high-power radar at the Ballistic Missile Early Warning site at Clear, and that might set off the explosive. Tom was able to reassure the captain that (a) the actual amount of explosive was very small, in spite of the overall weight of the package; (b) the igniters had been shorted out prior to packing and were in a metal can safe from the radio frequency energy from the radar; and (c) he felt comfortable enough that he would be riding as a passenger on the flight to expedite their delivery to Poker Flat that evening.

Tom flew back from Anchorage to Fairbanks without incident and arrived at Poker Flat a little after midnight. Our launch officer, Carroll Coe, and the Goddard rocket motor men installed the initiator, and by 2 a.m. we were ready to launch. The aurora was there in the sky but not of the right type for this particular study, so we launched the next night.

Minus 40° Causes Failure

It had been −40° F and colder for several days and nights near the end of February 1974 before we launched a payload into aurora. The experiment was aboard a Paiute Tomahawk rocket from the HAD launcher on pad 2. Seconds after liftoff, Carroll Coe told us he did not think the Tomahawk would ignite, because he had not seen a blip on his firing-current ammeter that would tell him the rocket had received current.

Firing current travels simultaneously to the wiring for the Paiute and Tomahawk rockets but is held from reaching the firing delay train for the Tomahawk by the first motion switch at the aft end of the rocket launcher. That switch is held open before launch by the weight of the rocket and payload in the near-vertical launch position. As the rocket moved upward, that switch should have closed and delivered electrical current to a train of explosives that about 20 s later would ignite the Tomahawk propellant.

Carroll alerted our roadblock observers to look for the flare of second-stage Tomahawk ignition. They reported that they saw nothing. Tens of seconds later, we saw a bright flash on our range surveillance television systems. The Tomahawk blew up on impact atop a hill about two miles downrange from pad 2.

Carroll, along with Ed Butterfield and Jim Carver from the White Sands Missile Range support group for the DNA, went down to pad 2 to see if they could figure out why the Tomahawk had failed to ignite. We had accomplished several successful launches from this particular launcher but never at 40 below. They brought the switch box from the pad 2 launcher back to the blockhouse, opened it up, and found the grease in it had stiffened in the cold. The spring that moved the switch could not move it fast enough to close and send current to the second-stage Tomahawk. They cleaned that grease out of the mechanism and replaced it with lithium-based grease that we used on other low-temperature mechanical systems at Poker Flat. Following that event, we made sure to clean and put low-temperature grease in each first motion switch on each of the launchers at Poker Flat.

A few days later, Merritt Helfferich and I investigated the impact site. We hired a helicopter to haul us and two snowmachines to the ridgeline less than a mile from the impact site. We floundered about in deep soft snow with our snowmachines but could not reach the site. The next summer I hiked up and easily found the impact crater. It was no more than 3 feet deep in the shallow soil that covered the rocky ridgeline. I recovered only a few pieces of unburned Tomahawk propellant. The metal of the Tomahawk and the payload had disintegrated in the explosion of the Tomahawk propellant.

Apache Igniter

We launched a Paiute-Apache rocket motor combination with a payload developed by Jim Barcus of the University of Denver, launch number 38, at 12:12:00 on March 28, 1972. Everything came apart a few hundred feet up. We found the nozzle of the Apache rocket motor in the middle of the track into the Caribou Creek watershed the same day. We found the bent and broken payload on a bank along Chatanika River a few hundred yards from the launchpad the next day. But we were puzzled that we did not find the brilliant white Apache motor right away. It was about 6 inches in diameter and 10 feet long, and it seemed to have disappeared.

A couple of years later, while recovering something else along the Chatanika River with a helicopter, Beau Battey spotted our missing white Apache motor underwater and tucked under a dirt overhang. He put a line on it, hoisted it out with the helicopter, and brought it back to Poker Flat. The thrust of the Paiute had been so great it had put a smooth bend in the normally straight Apache.

I took the broken unfired igniter out of the Apache just to see what it looked like. The MRN team at Poker Flat was successful in getting the army at Fort Wainwright to dispose of the bent Apache for us a couple of weeks later. Much later, my curiosity got the better of me and I got the igniter out to look at it again. It was open at both ends and you could see through it. The igniter material appeared to be a layer about a quarter-inch thick on the inside of the metal outer shell. The outer shell looked like woven metal.

Seventy-Five Below Zero

I walked into the lunchroom at Poker Inn on a cold December afternoon and asked who was driving the Plymouth rental car. Dow Evelyn of Defense Nuclear Agency got a pained look on his face and said, "I am, what's wrong?" I replied, "The hood seems to be firmly underneath the rear end of our range truck." Everyone in the room got up and went outside to look.

Dow had come to watch the launch of a 31-inch Sergeant rocket motor, without a scientific payload, as a test. The temperature had been about 20 below when he arrived at Poker Flat. But before Space Data finished installing the fins on the Sergeant, the temperature had slid down to −40. We had learned not to operate our hydraulic equipment when temperatures went below −40, because the seals in the hydraulic cylinders failed. Our temperatures continued to drop and had been at a steady −75 Fahrenheit now for several days.

Dow had left his rental Plymouth parked behind the truck earlier in the day and turned off the ignition. About 2 p.m. he went out and started his car, but the automatic transmission would not move it. With the engine running, he had put the automatic transmission into what he thought was neutral, so the running engine could warm the transmission fluid. However, it appeared that he had accidentally put it in drive. Once the fluid started warming up, the car slowly crept forward, forcing the hood underneath the rear end of our truck.

Dow shifted the transmission to neutral and then several of us sat on the Plymouth's hood, lowering it enough that it no longer touched the iron underneath our truck. Dow put his transmission into reverse and eased his Plymouth out from under our truck. There was not a scratch to be seen.

Dow came back in and told us that we should reschedule the launch attempt to the January launch windows for other rockets. Maybe it would be warmer then.

In a total fit of stupidity, I decided to ignite it, reasoning that it would surely not burn for more than a fraction of a second and would have no thrust because flame could come uniformly out of both ends. I was standing on the ground outside the kitchen end of Poker Inn. As a precaution I put the igniter on the ground and stepped on it to hold it down. I lit the active material with a match and watched the flame grow and sparkle. Not being contained, it burned longer than I had anticipated and it was trying to move underfoot, indicating there was some disparity in the thrust coming out of one end versus the other. When smoke started coming out of the side of the igniter, I feared it would burn my foot and stepped away. Freed, the igniter became airborne and clattered once against the building, narrowly missing me before falling spent back to the ground. I was mortified that I had done such a dumb thing.

An Eclipsed Launch, 1972

I remember Hugh Anderson's voice from our Fort Yukon Observatory, 150 miles north of Poker Flat under the apogee of the proposed rocket launch, coming over the scientists' launch-decision intercom at Poker Flat. He was telling his co-principal investigator Paul Cloutier, who was at Poker Flat, "You are not going to believe this, but the moon is disappearing."

The launch crew and I were listening in from the blockhouse at Poker Flat and were incredulous. Hugh then said, "There is an eclipse of the moon underway." I remember him quietly cursing something like "for crying out loud." Hugh told me that the presence of the moon was a go/no-go mission success criterion for this launch and that by the time this almost 2-h eclipse was over, their launch window for that night would be over too. He told me to scrub the mission for tonight and announce the opening of the launch window for tomorrow night, which I did.

It turned out that neither the principal investigators nor the several graduate students who were with them for the launch had thought to check the atlas of eclipses. Hugh, Paul, and their graduate students had built one of the most complex and innovative 9-inch-diameter sounding-rocket payloads yet constructed for studies of the aurora. During analysis of data from past launches, they had been plagued with problems determining the aspect angle of the rocket payload with respect to three-dimensional space, the Earth's magnetic field lines, and aurora. They had flown a conventional Earth's magnetic field line aspect sensor on those payloads, but they had not achieved the precision they wanted to understand the data they were acquiring of electron density as a function of energy and pitch angle distributions with respect to the Earth's magnetic field lines and aurora.

They set their payload engineer Del Oehme to the task of designing and building a lunar aspect sensor for this payload. The sensor was quite simple: a pair of curved clear lines with a solid-state detector spaced behind them. The sensor was mounted in the skin of the rocket. The moon's position in the sky is known with great precision. By measuring the signal from the lunar aspect sensor, both the spin rate of the rocket and its aspect or orientation in space could be determined with great precision.

The light of the moon typically makes it harder to see aurora, and I remembered working with Hugh and Paul to determine when the moon would be in position in the sky for their lunar aspect sensor to work properly and still have them be able to see the aurora.

We had counted this rocket down every night for nearly 3 weeks. Now we commiserated and then gave up and laughed that when the aurora and launch conditions were perfect, a lunar eclipse got us. Sort of like Murphy's law: If anything can go wrong, it will.

Breakfast Chilis in New Mexico

In early fall of 1972, I took Geophysical Institute business manager Sue Johnson with me to White Sands Missile Range in New Mexico to negotiate the finances for future contract work for the meteorological rocket operations at Poker Flat.

Norm Byers of the Atmospheric Sciences Laboratory (ASL) at WSMR was our main contact for the army MRN program at Poker Flat. Norm looked like he had been out in the New Mexico sun too long. His eyes were crinkled and he chain-smoked cigarettes.

I also took Bob Spies from the GI electronic shop with us. We had heard that WSMR had adapted a small gun-laying radar for use in tracking wind-weighting balloons. Bob was a genius with great practical knowledge of all sorts of electronics. He also had experience with software. He could design and build almost anything. Bob helped us secure the loan of a small radar from WSMR for use at Poker Flat. The NASA radar at Poker Flat at this time was being used to track wind-weighting balloons and rockets. We hoped to lessen their workload and at the same time connect the output of the WSMR radar directly to the wind-weighting minicomputer we called FERD in the blockhouse at Poker Flat.

While we were there, Alton Duff showed Bob and me the Nike Hercules radar system they were using to acquire data on balloon and rocket launches at WSMR. I was aghast at what a kluged-together system it was. Electronic cabinets were open with wires dangling out connected to outboard high-voltage capacitors and jury-rigged components. I was cautioned to move carefully, and more than once I heard and saw high voltages arcing. The White Sands system was a test bed for ideas. Once they got things right at this facility, they put together good-looking, high-quality gear to send to Poker Flat. I learned that the system at Poker became their flagship, the best they had.

While negotiating the contract, we learned of a major experiment that would be done early the next morning involving observations of air warming and rising as the sun did. The science questions being investigated were how atmospheric turbulence develops as the sun rises and warms the still air of a typical White Sands morning.

We wanted to watch this, so we got up early and went to the restaurant in our motel about 1 a.m. The short-order cook made us large chili-laced omelets slathered with hot salsa.

By 5 a.m. my stomach was churning. Along with the black coffee that I drank, the omelet was creating an enormous amount of gas distress in my intestinal tract. Thankfully, I was able to stand far away from anyone else as we stood outside. I felt terrible but laughed to myself, because it reminded me of the cowboys sitting around eating a dinner of beans and farting in the movie *Blazing Saddles*. However, in the midst of my intestinal torment, I was watching one of the most beautiful sets of scientific experiments I had ever seen.

Bob and I watched sunlight reflect off a large, instrumented balloon hanging 100,000 feet above the Earth. As the sun rose, its light glinted off other balloons at intermediate altitudes. Some of these balloons had conventional radiosonde onboard to measure temperature, humidity, and wind direction and speed from the surface to 100,000 feet. We watched the launches of several small Loki-Dart rockets that deployed high-altitude dropsonde above 200,000 feet to measure wind speeds, direction, and temperature above balloon altitudes.

But the most spectacular of all were the several Nike Smoke rockets that were launched. These were single-stage Nike rockets with an approximately 7-foot-long nose cone bolted to their front ends. Inside was a pressure tank of dry nitrogen and a canister of liquefied titanium tetrachloride—the material that old-time skywriters used to write messages with their biplanes. Within a fraction of a second of liftoff, an electrically activated valve allowed the pressurized dry nitrogen to push a small stream of titanium tetrachloride out a hole in the side of the nose cone. When titanium tetrachloride meets with moisture in the atmosphere, it immediately turns into "smoke." The stuff is somewhat hazardous. It is corrosive, and the chemical reaction is exothermic. It gets warm. The Nike was pushing a relatively light load, perhaps 700 pounds, and easily made it up to about 30,000 feet at apogee.

The first Nike Smoke quickly arched up from Launch Complex 36, dispensing its trail. We were a couple of miles away. The whole trail remained in place like a gigantic St. Louis Gateway Arch for perhaps a minute before it was slowly torn apart by shear winds at several levels between the surface and 30,000 feet. It was magnificent to watch this rocket trail smoke and leave this arch in its 3- or 4-min journey. Equally as interesting was watching it terminate by slamming into the gently sloping foothills of the mountains a few miles away. We could easily hear the downrange impact in the stillness of the cool morning air. We watched several more launches from 3 a.m. through 8 a.m.

Sue and Bob flew back to Fairbanks, and Alton took me to dinner at La Posta in Las Cruces, New Mexico. I had some of the finest Mexican food there I have ever eaten. After we finished, we were given a glass of milk and sopapillas filled with honey, not so much as dessert but to counteract the spicy-hot food.

Antenna Fix

We were launching a Nike-Tomahawk on October 21, 1972. A fraction of a second after liftoff, the radiotelemetry signal from the rocket died. The Nike first-stage rocket burned perfectly, separated, arched over trailing a soft burn, and impacted a mile and a half away just as it was supposed to. Twenty seconds after liftoff, high overhead we watched the second-stage Tomahawk lift off and burn for the 8 s it was supposed to.

We walked down to pad 1 and found one of the two ejectable doors from the payload lying on the launchpad. The radiotelemetry should have been protected from aerodynamic heating by the door, but it had died before the rocket had reached an altitude where that could have caused it to fail.

Charlie "Soup" Campbell and Joe Modlin of Goddard Space Flight Center represented NASA sounding rocket construction on this payload. They quietly told us that perhaps the antennas had failed, because they had on a recent previous mission. These antennas, which operated in the 220 MHz range, stuck out from the sides of the payload at about a 30-degree swept-back angle. They operated in the 220 MHz range. Soup and Joe said that while the antenna design was one of long-standing and well-proven, a new contractor had provided these, and they had not yet undergone shake or vibration testing at NASA Goddard.

The payload had been built for principal investigator Larry Cahill of the University of Minnesota and his co-principal investigator Roger Arnoldy. Larry was a consummate gentleman, mild-mannered, and unflappable, with many NASA rocket launch successes and a few failures behind him. He politely asked for details, while Roger was slightly nervous all the time. But if looks could kill, it was the intensely angry and nervous graduate student Paul Kintner, whose PhD would in part come from data acquired from this launch, who was furious at only now learning about the antennas.

Soon after this presumed antenna failure, Jim Heppner from Goddard arrived at Poker Flat with no fewer than four Goddard-built payloads with identical antennas. Somehow this known potential antenna problem still existed and had not been caught until Jim and his payload crew were deployed and in the field with us at Poker Flat. Jim was almost as unflappable as Larry, since he was also a rocketeer with many launches under his belt. But Jim was also a pretty high-ranking Goddard scientist by this time, and he started raising hell in telephone calls with Goddard management four time zones to the east of Poker Flat.

In the midst of this, Lou Baim, our payload engineer, learned about the problem and stepped forward with a proposed solution. Lou took one of the antennas to the GI machine shop in Fairbanks and had the head of the machine shop, Jim Parry, design and carve a block of stock Teflon so that it fit snugly between about 4 inches of the swept-back antenna and the payload body. He reasoned that this would help stiffen the antennas and suppress the vibrations that had caused them to break off.

There was no way to test Lou's fix except to fly it. Jim and his payload engineer successfully argued Goddard management into approving using the field fix on at least one launch. We launched, and radiotelemetry was solid throughout the flight. Goddard approved adding the same fix to the other three payloads, and they too flew without failures.

Almost a year later, Larry, Roger, and Paul came back to Poker Flat with a new payload. It was launched March 16, 1973, and flew well and transmitted data about the aurora successfully throughout its flight.

Later still, NASA moved their sounding rocket radiotelemetry from 220 to 2200 MHz and tasked their radiotelemetry contractor, the Physical Science Laboratory of New Mexico State University, to design, build, and test an antenna that did not stick out from their payloads but wrapped around their circumference. They designed and tested this configuration for various payload diameters. The design was so successful it is still in use.

ICECAP

I was especially proud of an effort I initiated for the MRN at Poker Flat to support one of the later missions of the Defense Nuclear Agency's ICECAP, which was designed to see if infrared emissions of the aurora would mask the ability of space-borne infrared sensors to detect an enemy intercontinental ballistic missile coming over the horizon. The US Defense Department interests focused on whether an enemy ICBM beyond or in the aurora could be detected with sensors aboard satellites orbiting the Earth. We launched our first ICECAP rockets from Poker Flat in the spring of 1972.

Experiments involving targets and sensors carried by sounding rockets operated outside the atmosphere above Poker Flat, but there were ground-based observations too. Infrared emissions are absorbed and degraded as they pass from the upper atmosphere through the ozone layer and Earth's lower atmosphere to the ground. The amount of degradation was determined by computer modeling of the atmosphere with guesses about temperature, density, and humidity structures that the infrared radiation would pass through. Ultimately the ground-based signal was corrected by computer models to estimate what the real signal had been outside of the Earth's atmosphere over Poker Flat. When we launched a real infrared instrumented rocket, we could get a real number of actual emissions up high compared to what got to the ground.

I knew enough about the science involved to know that real-time knowledge of atmospheric structure derived from the balloon and the rocket-borne MRN program could be combined with ground-based observations to improve the computer modeling. I convinced the ICECAP folks to talk with the army MRN program folks at Poker Flat. The result was that ICECAP paid the army MRN people to get data with their balloon and rocket program when the ICECAP ground-based sensors were in operation. The ICECAP atmospheric computer modeling programs were greatly improved with a few coordinated observations.

DNA funded the Air Force Geophysics Laboratory (AFGL) of Hanscom Field, Massachusetts, to design and build state-of-the-art near-infrared sensors capable of operating from a sounding rocket. DNA also funded Stanford Research Institute to move its incoherent scatter radar system from Palo Alto, California, to its position a little over a mile south of Poker Flat.

Initially, AFGL built sensors for their rockets and flew them from Fort Churchill, Canada. Their first flights involved adapting hardware to fly successfully aboard rockets and were understandably engineering successes and unfortunately scientific failures. The first ICECAP flight from Poker Flat in the spring of 1973 was an unqualified success for AFGL's and DNA's efforts.

DNA immediately ramped up funding for all participants, including Poker Flat. During this buildup, federal defense-related agencies flew C-5A planeloads of equipment into Eielson Air Force Base some 50 road miles from Poker Flat and asked that it be delivered. Eielson is part of the Alaska Air Command (AAC), whose headquarters are in Anchorage. In the fall of 1974, six 40-foot-long air force tractor trailers carrying S-band telemetry equipment arrived at Poker Flat, along with two trailer-mounted S-band receiving dishes. The AFGL sent rocket motors, payloads, liquid nitrogen, and liquid helium through Eielson Air Force Base to Poker Flat, as well as several trailers of gaseous helium used to inflate wind-weighting balloons. Ray Wilton of AFGL made demands on how AAC would deploy its search and rescue helicopter assets from Eielson to recover parachuted rocket payloads. One effort by a navy lab required deploying a complete radio receiving unit to Burnt Mountain north of Fort Yukon. Burnt Mountain had a radioactive thermal generator to generate electric power, which made it a bit of a top-secret site. It was a site that transmitted seismic signals it picked up, presumably from Soviet Union nuclear weapon tests, to US monitors. AFGL worked with AAC to set up recovery of rocket payloads with the search-and-rescue group at Eielson Eielson Air Force Base that flew HH-3 helicopters and had them do some practice runs from Eielson to the middle of the White Mountain Range and land at Poker Flat on their way back to Eielson like they would if they recovered a payload. The Office of Navy Research sent a 40-foot instrumented van to Fort Yukon where they set up antennas to do radio work in conjunction with launches from Poker Flat.

The impact of all this on AAC was huge, and within a few weeks of the completion of the ICECAP program that spring they called Merritt Helfferich and me to their Alaska headquarters at Joint Base Elmendorf in Anchorage to explain what in the world was happening at a University of Alaska facility. They said it was urgent, so Merritt and I flew down the next day to meet with the commander and staff, who indicated that ICECAP had caused them problems. We were unaware there was a problem. Each government group had told us in advance what they were doing. Turns out the AAC had memorandums of understanding with all of the groups who were bringing their equipment for launches for Poker Flat, but they had no coordinator for the sum of their work with AAC that fall. Merritt and I alternated explaining and asking questions because we were not completely aware of how ICECAP had impacted AAC. I thank my lucky stars Merritt was with me, because between us we figured out what had happened.

Alaska Air Command told us that because the impact on them had been so large and relatively unexpected during the last rocket launch season, they wanted us to be the communication hub between user agencies and tell them as early as possible in the future what services they needed to be prepared to deliver. And they were going to insist to these many other federal agencies that their requests to AAC had to go through Poker Flat first and that Poker would coordinate their activities with AAC.

As it turned out, the Poker Flat users never impacted AAC again like they did in the fall of 1974. The telemetry equipment stayed at Poker Flat. The Office of Naval Research did take a C-141 to Fort Yukon to fly the trailer back to Eielson and then back to the Naval Research Laboratory near Washington, DC, the next summer, all without priorities. The HH-3 search-and-rescue group at Eielson Eielson Air Force Base recovered payloads from 1975 through 1989, when I retired from Poker Flat.

The 1973 Program

The 1973 launch schedule was extremely heavy. In the last 15 days of March alone, 11 big sounding rockets were launched. In August of that year, the MRN team supported a project spearheaded by Bill Stringer of the Geophysical Institute that sought to determine the correlation between noctilucent cloud occurrence and mesopheric temperatures as measured by sounding rockets.

One of the more interesting ICECAP missions was a coordinated program of rocket and balloon launches undertaken by AFGL in September 1973. Two 500,000-cubic-foot balloons carrying instruments to measure long wavelength infrared emissions to altitudes of 25 miles were released. Sounding rockets were also launched to obtain the same measurements up to above 62 miles, thus compiling a vertical profile of these infrared emissions. Also onboard the balloons were instruments to ascertain the presence of the so-called Murray effect, the scintillation of infrared signals that had previously been seen from high-altitude balloons. Results of that particular experiment were inconclusive, and the Murray effect was later proven, at Poker Flat in 1975, to be no more than atmospheric moisture condensation that results from the presence of liquid nitrogen used to cool the infrared detectors.

The second program that fall was done in conjunction with NASA's Skylab III astronauts. Barium shaped-charges were injected along the earth's magnetic field, and the resultant streak was photographed and observed by the astronauts as well as ground stations. This experiment involved a pair of payloads, both boosted by Black Brant rockets, and although the astronauts were unable to maneuver Skylab into a position from which they could observe the barium on the second shot, ground-gathered data indicated that the upper layers of the atmosphere were perhaps not as conducive as first thought but that particles might interact with and actually be accelerated up out of the upper atmosphere.

Over Our Shoulder Launch, 1973

The two-stage rocket motor combination of a Nike booster with a Tomahawk upper stage was one of the most reliable systems used to loft scientific payloads into aurora from the mid-1960s to the early 1980s. More than 65 were launched from Poker Flat during my 18-year tenure as director.

On February 6, 1973, we launched a Nike-Tomahawk for Karl Theobald of Los Alamos Scientific Laboratory. It carried state-of-the-art instrumentation to measure energetic ultraviolet emissions of the aurora. Something went wrong after the Nike burned out at 3.5 s into its flight and before the Tomahawk liftoff at 8 s into the flight. Radar and telemetry reported the payload was coming down. Internal timers ignited the Tomahawk at 8 s into flight, and it burned and thrust for 8 s, as it should. While there was nothing onboard the bare Tomahawk to tell us where it was going, the visible part of its burn was tracing a pattern in the sky indicating that it had pitched up and was going to fly "over our shoulders" to the southwest.

I came to Poker Flat with lots of science knowledge about aurora and very little knowledge about sounding rockets. I had worried about our unguided rockets, often launched at angles of 85° above the northern horizon, pitching back over our shoulder to Fairbanks, which was nearly due south. This was my first, and as it turned out only, experience of a rocket pitching over our shoulder in my 18 years managing the facility.

We had no way to document where the bare Tomahawk had flown and wanted in the worst way to know where it might land. The blockhouse telephone rang, and it was Hugh Anderson asking for me. Hugh and his wife, Wendy, lived in Walsh Hall on the UAF campus. From there he had observed the unexpected path of the Tomahawk burn and thought it landed somewhere near Nenana, some 50 miles southeast of Fairbanks. He knew something must have gone wrong for it to fly up and west of him.

Meanwhile, range radar and telemetry reported the payload had landed. Radar lost the ability to interrogate the radar beacon onboard the payload, but the telemetry signal was alive and well. Telemetry gave us a solid azimuth to the payload and radar gave us its best estimate of range before the signal disappeared. We ginned together a portable telemetry antenna and receiver and drove up and down the Steese Highway making direction measurements to triangulate on the payload. The payload transmitted for the next hour before its batteries ran out of energy.

We hired a fixed-wing airplane and flew over the area for several hours the next day but could not find the payload. As the years went by, we continued to look for that payload. In the summer of 2003, they did a controlled burn in the area where the payload should be, but no one spotted it then either.

The day after the launch, we contacted seismologists of the Geophysical Institute and asked them if they had any earthquake instrumentation that might pick up the crash of the 300-pound Tomahawk motor casing. They replied that they doubted it. They said their instrumentation was insensitive to impulsive noise.

We figured that the empty Tomahawk tube had probably tipped over on its side during its fall and came in, in a flat spin, and probably hit the ground at less than 200 miles per hour.

Launcher Accident

The spring 1974 launch season began with Jim Heppner's return to Poker Flat with three rockets. They were followed by five rockets in the ICECAP series. All five payloads, which were studying the physical chemistry of the aurorally disturbed atmosphere, were successful except one. Its cause was a combination of extremely cold temperatures—in the −50° range—and a 2-week-long wait for proper conditions. Cold congealed grease interfered with the rapid movement of the first motion switch used to fire the second-stage motor, and it failed to ignite.

On March 8 we loaded a 31-inch-diameter, 12,000-pound rocket on the MRL 7.5K launcher on pad 4. We were working in partnership with Los Alamos National Laboratory and NASA to illuminate the Earth's magnetic field lines from Alaska to the Southern Hemisphere with barium. The next morning, it was −40°.

They started loading the payload, which consisted of an upper-stage rocket motor of about 400 pounds and 350 pounds of high explosives with seven barium canisters, and batteries to set off the explosives. Carroll Coe, Ivar Stevens, and Walt Suiter were on pad 4. Walt was from Sandia Corporation (now Sandia National Laboratories) in Albuquerque, New Mexico, and was supporting the joint LANL/GI mission. Walt was in charge of the operation, and Carroll and Ivar were helping. No one was wearing hard hats.

As they were using chain falls on the launcher boom to lift the upper stage and payload up onto the launcher to mate with the first-stage rocket motor, the jackscrew driveline, which held the boom, came apart and the launcher fell. Walt was hit on the head by a launcher boom. He was very bloody but talking up a storm. He told us his right leg and right jaw were broken, and he had a cut on his head.

They got Walt off the launchpad on a stretcher. Carroll went to his old International Travelall station wagon, threw a bunch of junk out, and got Walt in. They met an ambulance at the top of Cleary Summit, and 45 min later Walt was in the Fairbanks hospital.

I visited Walt in the hospital. His right leg was broken above the knee with a compression break with lots of splinters, and he had a hairline crack in his skull (mild concussion), a surface wound on his right cheek, a bad bruise, cut skin, and a black eye. He said they held off setting his leg because of crud in his lungs. He had a cold when he got hurt and that complicated things. He was doped up for the first week or 2. I visited him each day, and there was always a different woman in his room. I thought they were girlfriends. One day he asked me who they were, and so I asked them. Turns out they were part of an early-morning prayer group. They had read about his accident in the paper and selected him as the target of their prayer.

I learned from Walt that he had nearly been killed in a freak accident during World War II. He had been training to fly in a B-24 bomber in Florida. The plane taxied out onto the runway and waited for morning fog to dissipate. While waiting, he and others dropped out through the bomb bay to have a smoke. When the fog started to clear, he was the last back in and did not make it from the bomb bay through the wall into the cockpit before takeoff. On takeoff, the tip of a wing hit an upright crane near the end of the runway and down they came. Everyone was killed except him.

Walt sued the university and the company that made the launcher for $375,000 to recover his medical costs because his employer, Sandia, wouldn't pay. A year later I talked with Walt, and he said that they wrapped plates around the splintered area of his upper femur and squeezed it back together. He said it grew back together but was slightly crooked. His right foot was at an odd angle, and to fix it they would have to break it again. He did not do that, but he was able to play tennis later in his life.

After Walt's accident, we tried to figure out how in the world we were going to make the mess on pad 4 safe. The high explosives were still there and all tangled up. My first opinion was to write it all off: blow up the pad, rocket launcher, etc. so no one would get hurt. The only thing that kept me from following through was that no one could assure me it would completely get rid of all of the high explosives. It might get rid of 95%, but then a chunk of unexploded high explosive might be thrown out and end up in the middle of the Steese Highway, like a booby trap for some unsuspecting person to set off.

So instead of taking any action, we waited for 5 h for it to "cook off." In other words, if anything was getting ready to blow up by itself, let it. Then we very carefully inspected the mess before dark, ascertained it looked safe, and then disarmed a few of the igniters that we could reach without jiggling anything.

We had spare barium liners for the explosives. We took two out of their boxes and placed them inside an open-air box with a loose canvas over them. Normally the payload surrounding the barium liners was pressurized with dry nitrogen gas. The cover plate in the payload nose cone for the barium liners had opened up, releasing the dry nitrogen. Barium will oxidize quickly in moist air and could actually burst into flames. Fortunately, because it was so cold, the air was dry. Over the next 3 or 4 days we saw no sign of a problem with these liners. As I recall they were outside near the blockhouse.

Next we called high explosives experts at Los Alamos. They told us that as little as a quarter-inch slip in broken high explosives would set it off. Fortunately that never happened.

On Monday, we got the rocket and payload safely out from under the launcher in about 18 hours of careful and slow work.

Later on, we learned that four 1¾-inch bolts suddenly broke, likely because of the cold temperatures and heavy weights. They were handmade bolts and had a sharp transition from shank to head. Sharp edges create strains in the metal. They came this way from the factory. We also know that seven people owe their lives to an oddball hunk of steel we had welded onto the launcher to protect umbilicals from the rear turret to the launcher boom. That piece kept the launcher boom from crushing the high explosive payload.

Besides recovering from that launcher accident and determining why it occurred, we loaded and shot two rockets and safely rigged another launcher so we could shoot the second big rocket. We still had four rockets to shoot by Wednesday night and another five to rig and shoot during the second or third week of April.

We worked for 2 and a half weeks, got the explosive payload out from under the launcher, and then dealt with other problems 20 hours per day. Then we launched the second big barium shot (same as the one that the launcher fell on). At 33.2 s into flight, the first-stage motor blew up, scattering high explosives and toxic barium all over the ground 10 miles north of Poker Flat. That was the end; they took down the remaining rockets and headed for home. It was a bust.

The First 20,000-Pound Launcher

Soon after we recovered from the spring 1974 failure of an MRL 7.5K launcher on pad 4 and that spring's busy launching season, our DNA program manager Herb Mitchell called to discuss bringing a MRL 20K launcher to handle larger-diameter rocket motors that were on his schedule for Poker Flat.

By the spring of 1974, Poker Flat had three MRL 7.5K launchers on pads 1, 2, and 4, and one Sandia HAD launcher on pad 3. All three of the MRL 7.5K launchers were on 24-hour recall as part of the Operational Safeguard Program (described in Chap. 2).

Herb described how there had once been a total failure of a MRL 20K launcher during the launch of a large rocket from Eglin Air Force Base in Florida many years before. The rocket hung off to one side and when launched, the azimuth brake on the launcher proved inadequate. Not only did the rocket fly off course; its flight was such a danger to people and property that a command destruct system was used to blow it and its very expensive payload to such small pieces they would not hurt anyone as they fell back to the Earth. The launcher itself had also effectively been destroyed in the launch process.

Launch officer Carroll Coe gave me a picture of the damaged Eglin MRL 20K launcher. Hanging beneath the launch rail was a large structure built to hold a 30-inch-plus diameter rocket motor and payload. This was equipped with bearings so that before the rocket was launched, it could be spun up with electrically driven motors to develop the angular momentum that would help the rocket better fly the desired trajectory. Thrust from the main rocket motor was directed onto a large flat surface that was a part of the system that allowed the rocket to be spun up.

The launch boom of a MRL 20K launcher hangs off to the side of the upright pedestal, which is 8 feet in diameter. The thrust of the rocket motor on the flat mounting plate during that launch at Eglin caused the boom to rotate, and the rocket went off-course. Electric-motor-driven gearboxes are used to set launch azimuth and elevation. The azimuth drive system is mounted on the upper inside wall of the pedestal. The torque during the launch bent the azimuth engaging gear shaft system away from the azimuth ring gear, so the launcher was free to move. The MRL 20K had no other azimuth brake mechanism beyond this azimuth gear drive system. So one of the bigger problems I had to solve in installing the 20K launcher at Poker Flat was to find a way to keep the launcher from rotating in azimuth during launch.

During my annual trip to NASA Wallops Flight Facility in early summer 1975, my good friend Jim Gray took me to an area for surplus materials. I asked if we could have a large-diameter 12-foot-long launcher boom extension lying there and an intriguing 40-foot launch van. Herb paid for the launcher boom extension to be sent to Ed Butterfield at White Sands Missile Range and for the launch van to be sent directly to Poker Flat.

We were using a standard 36-foot-long launch rail assembly, which typically gave us 24 feet of guidance for most first-stage rocket motors. Launch rails are adjustable so that you can have simultaneous drop away from the launch rail of the launch lugs of the first-stage motor. The MRL 20K boom was only 20 feet long, and the launcher boom extension I got from Wallops would add another 12 feet.

The 40-foot van, when the rocket launcher inside was removed, would provide a heated and lighted workspace for rockets and payloads installed on the MRL 20K. The narrow roof opening was an almost perfect fit for the 20K launcher and extension boom. We would do the modifications at Poker Flat.

I learned that this van had been developed to launch a rocket-borne broadcast communication platform. Only a few of these vans were built, and as far as I could tell, none had ever been deployed. Special launch sites had been built for them in the Midwest. These vans were meant to move around and not stay in any one place for long, so the enemy could not find and destroy them.

Ed told me his people would mount the Wallops boom extension on the MRL 20K and he would mount an Honest John launch boom under the 20K boom and boom extension, and then he would install the standard launch rail under that. When I visited Ed at WSMR, I saw the Honest John launch boom, which looked horrendously heavy. It consisted of a very thick I-beam about 10 inches square, heavily reinforced and 30-plus feet long.

I came back to Alaska in early July with two major MRL 20K installation problems to solve. The first was to figure out how to make an adapter to connect the 9-foot-diameter base ring of the 20K pedestal to the 6-foot-diameter circle of bolts embedded in the concrete base of pad 4 for the previous MRL 7.5K launcher. The second problem was the azimuth brake, for which something really stout would be needed.

The outer ring would need to be made from 2-inch-thick steel. Jim Parry of the GI machine shop said our best bet was to have it fabricated in a shop he had used in Tacoma, Washington. This ring needed to have holes that mated with the existing MRL 7.5K bolts in the concrete pad and threaded holes with 8-inch-tall, 2-inch-diameter bolts around the outer rim to match the holes in the pedestal base of the 20K. Jim also suggested that we have them drill a number of equally spaced 2-inch-diameter holes around the rim for more fasteners to the concrete in the ground. These holes would also have to be placed to avoid reinforcing-rod bars already embedded in the concrete of the launchpad. The location of the reinforcing-rod bars were identified in the original drawings for the construction of the base for the MRL 7.5K launcher.

Jim suggested that after we installed the adapter ring, we bore through the 2-inch-diameter outer holes in the ring down 3 feet deep into the concrete base pad and install ThunderStud bolts. The ThunderStud bolts, which were roughly 2.5 feet long and just under 2 inches in diameter, slid easily into the hole, but the mechanism on their base expanded to give a good grip on the concrete as we tightened them.

The shop in Tacoma accepted the order, fabricated the ring according to my design, built a shipping container that tipped the ring up so it had an 8-foot footprint, and shipped it by truck to Alaska. After we set the ring in place, we rented a large magnetic-mount electrically powered drill, and Lou Baim drilled the holes in the concrete and set and tightened the ThunderStuds.

After giving the whole installation some thought, we realized that when the MRL 20K launcher was elevated to launch angles, it was going to hit the pad 4 concrete miniblockhouse. A significant amount of reinforced concrete had to be removed. We decided to weld a custom solid-steel plate fixture over the gaping hole. We managed to get this project done before significant cold weather set in.

Because the rockets hung from the sidearm of the MRL 20K and would be more powerful, new, heavier steel blast plates would have to be welded down on the launchpad. We cut and welded three 4-by-8-foot, 2-inch-thick steel plates in a rough arc on the launchpad and bolted them in place.

In mid-February of 1975, Ed Butterfield's group put the MRL 20K launcher his group had modified on a truck for Alaska. The driver called each day to tell us how far north he had gotten. One day the driver said that he had been told he could not drive across the Shaw Creek Bridge, about 70 miles west of Fairbanks on the Richardson Highway. He would be held up for several days while emergency repairs were made to the bridge, which had been damaged by a tall truck load that had jammed into its structure.

The next day he suddenly arrived at Poker Flat. The State of Alaska had bulldozed a ramp down to Shaw Creek and back up again, and the ice on Shaw Creek was strong and thick enough to drive his heavy load across. We were all a bit incredulous that this Lower 48 driver hauling our almost-irreplaceable rocket launcher had taken this risk, but here he was, and our new launcher sat on his truck.

We rented a large crane and operator to install the MRL 20K launcher on pad 4. The temperature hovered around zero as the wind blew and it snowed. Eldon Thompson organized the installation. Eldon enlisted Lou and Soup Campbell to help him. I remember seeing Soup sitting inside the upper part of the pedestal as the sidearm boom of the launcher was brought into position, hanging from the crane. Soup successfully directed the crane operator so that he could install first one bolt and nut to hold the boom onto the pedestal and then another. He soon had all 13 bolts tightened to specifications.

Meanwhile, I had been thinking for the last 6 months about how we could fabricate an azimuth brake. I thought about this launcher brake problem for months. I sought advice from others. I led group discussions trying to find a plausible solution. Despite these efforts, the problem was unresolved less than 2 months before we were scheduled to use this launcher for a very important project.

One night a few days after we got the launcher installed on pad 4, I remembered the brake I had photographed while on a hike along the rim of the Snake River canyon in 1960.

The Snake River was a natural path for grain grown and harvested in the Palouse country around Pullman, Washington, to be transported to market in Portland, Oregon. But the Palouse country farmlands were nearly 1800 feet in elevation above the river. Early ranchers had a dickens of a time getting their grain down to the river. Around 1900, a group built a cable system that ran from the Snake River canyon edge down to the river and back up again. They hung individual sacks of grain on the cable, and gravity would move them down. They loaded commercial goods they wanted to haul up on the other cable. To control the speed, they built a huge brake band around the large upper pivot wheel you see in the picture on the next page. A large band of metal with replaceable wood brake pads was cinched onto the pivot wheel as needed to control motion. In the picture it is dangling below the wheel it brakes.

I found the 35-mm color slide I had of this brake, and first thing that morning, I set my projector up in the Poker Flat conference room and showed the crew my solution. I proposed we build an external brake assembly on the part of the MRL 20K pedestal that gripped the fixed pedestal when activated. I proposed that we build a big steel band, line it with belting, and use a small portable electric hydraulic pump to power a hydraulic piston to cinch the band to the pedestal. The response was immediate and positive. We did a quick calculation that showed us a brake like this would be more than adequate to restrain the launch forces expected. It was a simple to construct solution, and there were no engineering drawings.

Neal Brown's photo of the historic Snake River cable brake, used for moving sacks of grain

The Poker Flat version of a brake band. Photo by Dan Osborne

Eldon fully embraced the idea and construction started. The system was exposed to snow and rain, so they added a cover. Within a few days, we had the brake operational. We did not do any testing but were confident that it was going to work great. We activated the brake from the blockhouse by turning on and off an electric-powered portable hydraulic pump system in the pedestal.

We created an immensely powerful brake that has never failed to work properly and hold the launcher in alignment during the launches of various kinds of sounding rockets. Some years later, Space Data Corporation was hired to prepare two MRL 20K launchers for use at Kwajalein Atoll. They duplicated our design with success.

The Oosik Mission

One fine spring day in 1973, I received a telephone call from Milt Peek at Los Alamos National Laboratory. His management had figured out what an *oosik* was, and were near apoplectic that it was the code name for a rocket launch in a refereed journal article about to be published. When they said it was not an approved Los Alamos code name, he told them that it was a University of Alaska Geophysical Institute project he had partnered with.

Oosik was a hugely successful project involving the rocket, explosives, and a shaped charge of barium onboard rocket payloads launched from Poker Flat, driven by Neil Davis and Gene Wescott at the Geophysical Institute. We had launched it in March of 1972. Ted Lehne, newscaster and owner of Channel 11 TV and a radio station in Fairbanks, broadcast the launch live.

At the time, it had been postulated that the rocket might create a magnificent aurora, and indeed an aurora suddenly filled the sky with color and motion for the next 10 min. As it turned out, this was apparently a fluke. We were never successful again, and theoretical work soon came out stating we never could have been.

I assured Milt that I would call him back with a set of plausible science words for the acronym OOSIK within the next hour. After I hung up, I turned to Gil Moore and explained my dilemma. Gil asked how we had come up with that name to begin with. I told him that a couple of years before, Milt and I had worked several nights together in the Ester Dome Observatory west of Fairbanks waiting for an aurora rocket to be launched from Poker Flat. Milt had said, "You know, Neal, if we do create a shaped barium charge and fly it aboard a rocket on this proposed project, what do you suppose we should call it?"

I was at the time a graduate student with no connection with the rocket work, but I replied without hesitation, "Oosik." I told Milt with a smile on my face that it would be like a giant ejaculation of barium in the skies over Alaska, and *oosik* is the name of the penis bone of a walrus. We laughed, but as time went by and the project came to be, the name Oosik just stuck.

Gil was a good friend with many years of rocketry experience: he was at White Sands when some of the first German V-2 rockets captured at the end of World War II were assembled and flown. He worked for Thiokol rocket motor company and was now the head of a division he created called Astromet. He was on a first-name basis with the head of NASA and several astronauts. He knew how to create acronyms.

Gil walked about 30 feet through the hallway to the door, turned, and with a big smile on his face said, "Optical Observation of Sudden Induced Kinetics." I picked up the telephone and tried to reach Neil Davis, but he was unavailable. I called his co-principal investigator on this project, Gene Wescott. Gene answered, and I briefly explained our dilemma and Gil's solution. Gene said it was fine with him. I called Milt and told him what we had come up with and Gene's approval. Milt said, "Thanks, I think this will do it."

In 1978 Keith Mather asked me to write the feature article for the 1979 Geophysical Institute annual report celebrating the tenth anniversary of the first launch from Poker Flat. Peggy Dace Boyd was working for me and assembled our submission from what little I had written.

While Keith was editing our submission, he called Neil into his office, and they had a good laugh about how the words describing the acronym came about. It had been a surprising revelation to Keith, and he took it well since everything had come out fine in the end.

Ode to an Oosik
by Robert W. Service

Strange things have been done in the Midnight Sun,
and the story books are full—
But the strangest tale concerns the male,
magnificent walrus bull!
I know it's rude, quite common and crude,
Perhaps it is grossly unkind;
But with first glance at least, this bewhiskered beast,
is as ugly in front as behind.
Look once again, take a second look—then
you'll see he's not ugly or vile—
There's a hint of a grin, in that blubbery chin—
and the eyes have a shy secret smile.
How can this be, this clandestine glee
that exudes from the walrus like music?
He knows, there inside, beneath blubber and hide
lies a splendid contrivance—the Oosik!
"Oosik" you say—and quite well you may,
I'll explain if you keep it between us;
In the simplest truth, though rather uncouth
"Oosik" is, in fact, his penis!
Now the size alone of this walrus bone,

would indeed arouse envious thinking—
It is also a fact, documented and backed,
There is never a softening or shrinking!
This, then, is why the smile is so sly,
the walrus is rightfully proud.
Though the climate is frigid, the walrus is rigid,
Pray, why, is not man so endowed?
Added to this, is a smile you might miss—
Though the bull is entitled to bow—
The one to out-smile our bull by a mile
is the satisfied walrus cow!

Grok generated image

Barium Launches, 1973

NASA Wallops had been launching sounding rockets to study wind speeds, directions, and temperatures at altitudes of 25 to 55 miles from Point Barrow, Alaska, since 1965. They used a unique launcher for small Nike-boosted rockets called the Rocket Aft Grab, or RAG, launcher, built by Thiokol. We called it a zero-length launcher because once ignited, the rocket had only about 1 inch of guidance before it left the launcher.

In late fall of 1972, Jim Gray of Wallops called to tell us they were ending their launch program at Point Barrow and that they would make the RAG launcher and DOVAP tracking system that they had been using at Point Barrow available for us at Poker Flat. NASA shipped the RAG launcher from Point Barrow to Poker Flat.

We were in the midst of launching rockets in February of 1973 when the RAG launcher arrived. We were launching meteorological rockets for the army's MRN from a small metal building and concrete pad about 300 feet east of the blockhouse. We had firing lines and communication cables in place to this building.

The RAG launcher was designed to be portable and had three 20-foot-long legs arrayed in a flattened tripod. Eldon Thompson declared that we should set it on the ground near the MRN launchpad, level it, and freeze it in place to the ground with a slurry of woodchips frozen in water ice, which we did.

We were experienced in keeping rockets warm with Herman Nelson heaters while loading them on launchers, but working on the barium payloads while they were being installed on the rockets would require more protection. We had an empty military surplus fifth-wheel trailer van and a truck tractor to move it. The van was only 10 to 13 feet long. We added another 13 feet of roof to the open back end of the van and hung canvas curtains from the roof to the ground.

NASA installed an identical RAG launcher at Barter Island, Alaska. In March of 1973, we launched five rockets for NASA's Goddard Space Flight Center, and nearly simultaneously Gil Moore and his team launched five rockets from Barter Island. This was the beginning of the overall experiment to measure electric fields and neutral winds in the Earth's high upper atmosphere. Each Nike-Tomahawk rocket payload carried four canisters filled with barium powder and copper oxide, separated by a thin, fragile disk. An onboard timer fired a small charge to break each disk at appropriate times during the flight. Each rocket flew for about 6 min and laid a trail of four large puffs of barium along a magnetic south-to-north path. The near-simultaneous launch sequence from Poker Flat and Barter Island produced eight puffs of barium over roughly 10° of subpolar and polar sunlit ionospheric latitudes.

Each puff of barium first appeared as a puff of white about the size of the full moon. Within seconds, sunlight ionized the barium into uniquely blue-green ions that followed the Earth's magnetic field lines. Ground-based cameras were used to photograph the positions of the luminous barium clouds against night-sky star fields. Later, analysis of the position of the barium cloud and the associated cloud of barium oxide would be used to measure the electric fields and neutral winds, respectively, as a function of latitude. It was a well-thought-out experiment that I admired.

A few days after the March launches, I received a telephone call from a gentleman in Wrangell, in Southeast Alaska. He told me that the electrical power for his community had failed that night and that it had been his responsibility to drive some 15 miles south to the hydroelectric system to find the problem and get electricity going again. He had just gotten out of his car when out of the corner of his eye he saw something appear low on the northern horizon. While he was trying to figure out what he was seeing, another object appeared and he ducked, thinking something was shooting at him. He said the next appearance worried him even more, and by the time he saw a fourth something appear in the northern sky, he had convinced himself that space aliens were coming.

He said the things he had seen in the sky seemed to slowly fade away and that he regained his composure enough to get the hydroelectric generator system working again. He drove home mystified but reluctant to tell anyone else lest they think it a bit crazy. Just before calling me, he had read an article in the *Juneau Empire* newspaper about the visible chemical releases that had been launched that same night. He called me to learn more about the rocket-borne experiments and said that he could now tell other people that he had seen them. He was not sure he would tell his friends the rest of the story about what he had first thought!

First Echo Rocket Lands in Canada

In April of 1974, I stood not far from Jack Winckler, watching him inspect data in real time from his rocket payload that we had just launched. He waved me closer and pointed to significantly structured wiggles on a streaming chart record. He told me they were records of a beam of electrons, sent upward 1.5 s earlier from his payload, clearly bouncing back from the Southern Hemisphere and being detected on the same payload.

It was to me the most exciting scientific experiment ever launched aboard a sounding rocket from Poker Flat. It was the first successful experiment confirming how Mother Nature creates nearly identical forms of aurora in both the Southern and Northern Hemisphere at nearly the same time.

Jack, of the University of Minnesota, was the principal investigator for the aptly named Echo project. This launch was his first major success, and several more followed. His earlier Echo flights at White Sands in New Mexico and Fort Churchill in Canada had not been successful, but each had improved the engineering technology required for the success of this one.

Jack developed high-energy radiation instrumentation flown on high-altitude balloon payloads for measurements of processes in the aurora and then transitioned to rocket payloads. He was a consummate teacher and mentor to his many graduate students and staff from the University of Minnesota who came with him to prepare and launch his Echo rocket payloads from Poker Flat.

I became a sort of Jack Winckler groupie at American Geophysical Union (AGU) science meetings and always sought him out if he was there. His presentations and those of his students were always among the most interesting. They were clear, thoughtful, well-presented, and argued ideas backed up with the latest solid evidence. On more than one occasion, Jack sidled over to me saying, “Let’s you and I take a long walk and get something to eat.” We would amble along for 10 or 20 blocks discussing nothing in particular, enjoying the fresh air and scenery before finding a restaurant that we both agreed upon. We often spent so long talking we decided we needed to take a cab back to the AGU for the next presentations we were interested in.

The summer before the Echo launch at Poker Flat, Randi Wagner and I traveled to Minneapolis to meet with Jack and his team. There I learned that to meet his science objectives, Jack would need to launch east out of Poker Flat. The beam of electrons from his payload would travel over 65,000 miles out into space and back to reach their magnetically conjugate mirror point and then travel another 65,000 miles back along a similar path to Alaska, where if all went well they would be detected on the same rocket payload they had originated from. The beam was coded in a unique on-off sequence so he knew it was the beam we had originated. If everything worked right, the beam would return 1.5 s later, but it would return several miles magnetically east. So the rocket had to be launched to the east.

We had a can-do attitude at Poker Flat, with the science requirements always coming first over other range support requirements. We started conversations with the FAA to negotiate this kind of flight through the airspace they controlled, from the surface to 50,000 feet. We also started conversations with the Bureau of Land Management to find lands to our east that met the science requirements and would allow for the safe impact of the upper-stage rocket motor and payload well away from people. These negotiations took a long time but were ultimately successful. We had FAA clearance for flight and an impact point just west of the Taylor Highway that led from the Alcan Highway just east of Tok, Alaska, to Boundary, Alaska.

The rocket payload would carry a sophisticated high-pressure attitude-control system onboard. This system was serviced through stainless steel piping installed on the rocket launcher and connected through quick-disconnect couplings to the payload. The first major movement of the rocket would disconnect the couplings, and internal systems would seal off the lines so they did not leak.

A more difficult problem was that the first-stage rocket booster for Echo was predicted to land within a dozen yards east of the main electrical substation of Poker Flat. At this station the 34,500-volt transmission line from Fairbanks was reduced to 4160 volts for on-range use. If the booster fell on the power line, it could knock out power to the rocket payload telemetry receiving ground station. A possible alternative would be to operate that part of Poker Flat on an emergency generator system. In the end, the possibility of a rocket impacting the power line was a risk everyone accepted, and an emergency generator was not used.

Jack was a successful and therefore powerful NASA scientist. While at White Sands, he had support of the NASA contractor Physical Science Laboratory (PSL) of the University of New Mexico at Las Cruces for telemetry. He insisted they come and support him at Poker Flat.

I knew telemetry folks at both Wallops and PSL and I heard Jack fought long and hard to get what showed up at Poker Flat. Officially, NASA Wallops was to provide telemetry for the flight and PSL was to provide telemetry support only for launchpad check-out. But everyone knew that Jack had arranged with PSL for them to bring equipment adequate to receive data throughout the flight as well. A minor difference to Jack and PSL was that NASA had automatic tracking receiving antennas, while PSL would have to steer their receiving antennas manually.

I never did figure out why Jack was so critical of Wallops, but he said in vague terms they had upset him in the past with failed promises of support and he was going to make sure they never did again. But Wallops did succeed in acquiring excellent data on the first Echo launch from Poker Flat, as did PSL. On subsequent Echo missions, PSL showed up with less equipment and people that they used on lower range next to the blockhouse, payload assembly, and the launchpads at Poker Flat. Wallops provided excellent primary flight telemetry data acquisition, but PSL still provided launchpad check-out telemetry support and minor flight telemetry support.

Jack had many balloon and rocket experiment successes under his belt. He spent hours on the telephone talking with other scientists who were acquiring real-time data from satellites or ground stations surrounding and at Poker Flat. It was a real pleasure to work with someone so knowledgeable, thorough, and decisive.

Finally, everything was ready and conditions were right. For the first time, the countdown went below the stopping points of 5 min, then 2 min, then 1 min, and we knew that while we could stop the launch by Carroll not pressing his launch button at T-minus a fraction of a second, he would not have to.

With a mighty roar the first rocket motor stage thrust the second rocket motor stage and payload into the sky and fell back to Earth. Our electrical power remained steady after first-stage impact, so it must have missed the power line. So far, so good.

NASA radar reported that the rocket appeared to be flying at a significantly lower quadrant elevation (angle) than predicted. If the rocket did not fly high enough, the electron beam would be absorbed by the Earth's atmosphere and not escape with enough energy to reach the Southern Hemisphere and bounce back.

I had watched liftoff with Jack and the PSL crew next to their equipment parked at the Payload Assembly Building a good 1000 feet from the launchpad. We had specifically asked radar to report altitude of the payload above Earth as it flew, over a separate channel on the range intercom, and they were doing so.

We watched the PSL telemetry chart recording show that the attitude-control system had aligned the payload with the Earth's magnetic field lines as it flew and that the electron gun was working perfectly firing a beam upward, but there was no sign of a return beam. We wondered how much higher the payload would get at this new angle of flight.

Then the payload receiver started showing a weak return signal that grew stronger and stronger. In the 6-min flight, Jack had just a minute near apogee of clear, strong evidence that everything was working perfectly technically and scientifically. When first telemetry and then radar announced loss of signal, Jack lifted his eyes from the chart recordings and congratulated everyone around him.

However, radar reported the first stage and payload had gone far beyond the predicted safe impact point this side of the Taylor Highway in Alaska. According to their data, it had overflown the highway and landed in Canada. It had flown low because the disconnect system for the stainless steel attitude-control system on the launcher had failed to work.

Of considerably less import but of great interest to me was that the first-stage rocket had landed a scant two or three yards this side of the power line.

I asked Merritt Helfferich to tackle the Canadian impact problem. He set out to try and find out where it might be and to talk with the appropriate authorities. He determined that the area it landed in was remote from almost everything associated with human activities and talked with Canadian Yukon Territory officials such as parks, recreation, and land management. All found what we had done to be of little interest as it posed no threat to anyone. It seemed to me they were almost not curious. I joked with Merritt and few others that we had just returned the Canadian-built Black Brant rocket motor to Canada.

While Merritt had acted quickly locally, NASA administrators on the East Coast were going ballistic as soon as they heard one of their rockets launched from Poker Flat had landed in Canada. Their standard procedures, which had little to do with us, were that if there was any possibility that one of their rockets would land in Canada, they would clear such a possibility through diplomatic channels between the US government and the Canadian embassy in Washington, DC, months in advance. Apparently this sort of procedure was in place for NASA launches from the Michigan Upper Peninsula many years previously. Merritt worked with Neil Davis to keep NASA administrators from overreacting. They did fuss and fume, but not so much at us as at their own people for not anticipating this kind of problem. We heard that NASA headquarters did go through official channels to tell Canada what had happened. They did so with solid information that Merritt had developed about the Yukon Territory government people he had talked to, their telephone numbers and addresses, and their reactions. The tempest seemed to die a natural death, and we never heard about it again.

As it turned out, we never launched to the east again. Subsequent Echo-type payloads employed subpayloads ejected from a mother payload as it flew and measured other relevant parameters associated with conjugate aurora.

Twenty years later, co-principal investigator Tom Hallinan of the Geophysical Institute published new findings from the launch. He had flown onboard a research aircraft in the Southern Hemisphere and with a low-light-level television recorded the pencil-like beam of aurora the first Echo payload created in the Southern Hemisphere. An identical ground-based low-light-level television system in Alaska did not appear visually to record the return electron beam spot of aurora in the Northern Hemisphere. But two decades later, Tom replayed the Alaska ground-based data from magnetic tapes and, using sophisticated computerized television image techniques on a lowly IBM PC, discovered the return beam aurora spots.

Overflight of Arctic Village

We had just launched a rocket-borne experiment for the Ballistic Missile Agency, aimed far west of the town of Arctic Village. Our first visitor from the local BLM office had joined us in the blockhouse that evening to learn more about how we used the lands under multiuse land impact permits for the rockets we launched. Most of our rockets, including this one, were launched to impact on land controlled by BLM in cooperation with ten other federal and state agencies and Alaska Native corporations.

As soon as the rocket lifted off, I walked about 10 feet from the launch control console to the wind-weighting console in the blockhouse and watched the computer screens with the first real-time tracking and impact-point prediction software that Mike Cogan had developed. He had made two plots that were built on real-time radar tracking data. One followed the flight azimuth of the rocket as it took wind corrections, and the other followed its ballistic trajectory up into space and updated the downrange impact distance in real time on both plots.

As the rocket flew, the flight azimuth was swinging steadily east, but it did not stop west of Arctic Village where it was supposed to. Instead, it continued steadily swinging east until, to my horror, its movement slowed to stop right on the azimuth between Poker Flat and Arctic Village. The rocket was by now outside of the Earth's atmosphere and moving fast.

There seemed no way it would not land in Arctic Village.

It had been our policy for years never to even overfly any Alaska community. But the data now said we were going to land in one. Wind-weighting a rocket involves mathematically folding in rocket performance information from rocket specialists with the local wind speed and direction data as a function of altitude gained locally just before launch so the launcher can be aimed to land the rocket and payload in a place far from people. I could hear the BLM representative joking with people in the launch control room and was glad he did not know about the problem I saw looming on our computer screens.

I knew what Mike had done worked correctly. He had tested and proven it to me many times using radar tracking tapes from prior missions, although this was the first mission to use the data to predict in real time the actual flight of a vehicle. We had carried this real-time plotting effort out on our own, and few people and no agencies knew we were doing it.

As the rocket continued its upward arc, the downrange impact point remained firmly dead-center on downtown Arctic Village. As the rocket neared its highest point, or apogee, it continued to climb slightly and the predicted impact point started moving north of Arctic Village. It seemed impossible, but as each second went by after apogee, the impact point moved farther downrange. Before the NASA tracking radar from Poker Flat lost the payload radar beacon on the down leg at about 70,000 feet altitude and 350 miles downrange, the impact point was more than 100 miles north of Arctic Village at a position about halfway between the Brooks Range and the Arctic Ocean coastline.

Representatives of Space Data Corporation, who had provided us with the rocket performance data, stood with Mike and me in the wind-weighting room of the blockhouse wondering how this had come about. The BLM representative came in to congratulate us on a successful launch, told me he had had a great evening, and went home. I smiled and said nothing about what had taken place. I did not tell him that we had just overflown Arctic Village, which violated our internal rules and the fine print of our agreement with BLM and other agencies.

A day later I received a telephone call from Steve Fisher, the head of Space Data Corporation, telling me they screwed up and had entered the wrong drag performance data for the rocket configuration. They had put too high a number into their equations. They had cranked out some new numbers that fit with the actual impact point.

Steve and all of the Space Data crews had always been first-rate and to my knowledge had never led us astray. This was the first of three rockets, all of the same configuration, to be launched within a 2-week launch window. I thanked Steve and hoped the next one, now on the launch rail ready to go, and the one after that would hit their new predicted impact points. They did.

The rocket had landed in an unoccupied portion of the Arctic National Wildlife Refuge, and we notified the appropriate parties that we would search for and get it out of there. A couple of years later, we found and brought back the second-stage rocket and what remained of the payload from the Arctic National Wildlife Refuge.

I remember at one point communicating with Jack Parks, who headed up range safety for NASA Wallops, about some aspect of wind-weighting software. I heard that NASA Wallops had outsourced their software and did not know what was in it. NASA Wallops instead had a few test cases from the past, and each time the contractor revised their wind-weighting software, NASA Wallops ran those test cases. If the answers came out OK, all was good. If not, the software contractor was notified and made changes. But I learned that the contract limited the number of changes each year to something like 3 days of the software company's work. If more work was required, Wallops got hit with huge fees. At one point Wallops asked for a copy of our wind-weighting program, which we gave to them.

Chapter 7
Late 1970s

Catastrophic 100th Launch, 1975

The 100th launch from Poker Flat failed spectacularly because one rocket motor element had been left out too long at Elmendorf Air Force Base before it was warmed up and shipped from Anchorage to Fairbanks. We learned this only during our post-launch failure investigation. This launch occurred in the summer, which was both good and bad. Good because everyone was relaxed and enjoyed summer rather than winter weather; bad because of the potential to start a fire in the surrounding tundra. Tundra fires can burn for years after you think they are extinguished because they can smolder just below the surface.

The launch, which was carried out on the afternoon of June 13, 1975, was funded by Defense Nuclear Agency as a ballistic test of a Javelin-boosted Sergeant rocket. Edward Allen of Space Data Corporation was the principal investigator in charge of the launch.

We had installed a MRL 20K launcher on pad 4 the previous winter for 31-inch-diameter rockets. We were testing an air force surplus self-contained launch trailer for winter use and had taken the launch elevation screw system out of it. The trailer had internal explosion-proof wiring and was heavily insulated with blue foam in the walls. We loaded the rocket using this trailer and then pulled it some distance away, elevated the launcher, and prepared to launch.

The main rocket motor was a 31-inch diameter, 12,000-pound surplus US Army Sergeant. Two 9-inch-diameter, boost-assist rockets had been mounted on either side of the Sergeant. These had been shipped to us via the air force the previous winter. They were roughly 12 feet long and carried only a couple of inches of propellant lining them. At liftoff, each of the three rockets pushed with about 50,000 pounds of thrust. Each boost-assist motor burned for only 2 s through canted nozzles, so at liftoff all thrust went through the center of mass of the combination of all three motors.

N. Brown, *Northern Lights and Rocket Flights*, Springer Biographies,
https://doi.org/10.1007/978-3-032-14598-7_7

As luck would have it, this would be the launch recorded for *Lights in the Northern Sky*, a film funded by the National Science Foundation through a grant to Neil Davis at the Geophysical Institute to help people understand how we investigate the aurora with a variety of instruments, including sounding rockets. Greg Emmert of local public television station KUAC and UAF Media Services was on hand to document the launch with a 16-mm film camera.

Henry Cole was in the blockhouse wind-weighting, Randi Wagner was doing the countdown, and several of us were a hundred or more yards away, standing outside to watch the launch. I was near the Sandia launcher on pad 5. Greg was in position filming from just outside of the rocket storage building, a half-mile away and a hundred feet higher in elevation. We were all relaxed and confident that we were prepared during the few-minute countdown. We were communicating with the FAA, and all was well.

When Carroll Coe pushed the button to fire the three rockets, the two recruits and the Sergeant, it seemed like all hell broke loose. The Javelin on the right-hand side immediately exploded a fraction of a second after ignition, weakening the structure that held it, the other Javelin, and the four fins on the tail end of the Sergeant. I watched the whole event from perhaps 600 feet away. I saw the aft structure that held the fins peel off and fall away and immediately wondered where the rocket would go during its 40-s burn without fins to stabilize it. To my amazement, it kept going up. It seemed to wobble and start chasing its tail of flame, flying in an ever-rising circle as it disappeared into some patchy clouds about 12,000 feet above the ground. Telemetry and radar reported it was circling just above the clouds overhead, topping out at 18,000 feet. They reported that the parachuted dummy payload separated from the booster and was going to land less than two miles away, across the Chatanika River.

Dense smoke billowed from pad 4. We had Beau Battey sitting on the seat of an idling bulldozer. The air force trailer was burning furiously with a highly toxic black smoke. It was parked on a road, so we let it burn. But the tundra to the immediate south of pad 4 was also burning. The tundra was only inches thick on top of frozen icy ground, and Beau moved the bulldozer around the fire, avoiding the smoke and pushing burning tundra into a pile about 50 feet in diameter and a dozen feet tall. Beau and others spent the rest of that day making sure the fire stayed there.

Randi was in communication with fire suppression agencies to come help if the Poker Creek watershed caught fire. I remember sitting half in and half out of the pedestal for pad 5, telling Randi that it looked like we had everything under control, as we saw the payload come back down through the clouds and land across the highway. We had a helicopter standing by, both to make sure we could get to and fight a potential fire and to recover the payload.

We soon figured out that one of the boost-assist motors had exploded soon after ignition. We found the debris from its explosion on the launchpad. Much later, we learned that the propellant grain in that rocket had probably cracked a year or 2 before when personnel at Elmendorf Air Force Base in Anchorage had left it outside during below-freezing weather. It should have been kept around room temperature as per the transportation instructions. If frozen, the thin rocket propellant grain might have cracked, leading it to plug the exhaust nozzle and explode.

A few weeks later, I was given a short film called *A Day at Poker Flat* that started with the explosive launch, followed the rocket upward, and ended with the rocket chasing its tail before disappearing through the clouds. Made later, Neil's *Lights in the Northern Sky* ended with Greg's dramatic shot of the launch. I imagine most people who see the launch in the movie think it is spectacular and normal. I know people familiar with rockets immediately cringe and ask what went wrong.

Aurora in Stereo

One of my few published professional papers was "Altitude of Pulsating Aurora Determined by a New Instrumental Technique," coauthored with Neil Davis, Tom Hallinan, and Hans Stenbaek-Nielsen, published in *Geophysical Research Letters* in July 1976. My part of the "new instrumental technique" was to reinvigorate a long-standing unsuccessful project by providing a small television transmitter and receiver system so that the video picture from two low-light television cameras located a few miles apart could be seen in one location at the same time.

I had a great friend in DNA program management officer Herb Mitchell. He got us the money we needed to improve and operate Poker Flat for the ICECAP program for several years, in addition to negotiating and providing funding to the scientists and rocketeers for the ICECAP payloads they built and flew. One day Herb called to ask me if I had any use for a small television transmitter receiver system, which DNA had used to send television images from an unmanned drone airplane to a ground station many miles away. Within seconds I replied yes and told him about how our television group had for years tried and failed to do stereo imaging of the aurora. I was sure this equipment could help us be successful with that project.

Herb had the equipment sent to me at Poker Flat, and we set it up on a test bench, turned it on, and it worked. The transmitter was about 6 inches wide, 16 inches long, and 3 inches tall. One side was glued with heat-sink cement to an aluminum plate, and the other side had long fins that dissipated heat into the surrounding air. The receiver was much smaller. I bought a couple of 5 GHz feed horns, one for the transmitter and one for the receiver, and some power supplies that could be plugged into a regular electrical outlet.

It was not until near the end of the regular aurora season in early April of 1976 that we got the chance to put together a real-time aurora stereo system. One low-light-level television camera was located under a clear Plexiglas dome on top of the Ester Dome Observatory, 7 miles west of the Geophysical Institute. The other camera was under a similar dome on top of the Geophysical Institute. We had line of sight and a good radio-television signal between the two cameras. We electronically combined alternating images from both cameras into one output signal that went to both a tape recorder and a monitor at the GI site.

In the past, it had proved impossible to align the two television cameras so they saw the same aurora. At first, the two low-light-level aurora television cameras were some 27 miles apart, which was too far when looking at aurora only 60 miles above. Despite a lot of effort, no one could line up the images to determine the height of the aurora they were looking at.

Film photographers in the early 1900s had successfully measured the height of almost static aurora. Film in the 1900s would register an image of stars or aurora only when exposed for 5–10 min. Even then, it took quite a bit of preplanning to set the two precisely aligned cameras at two stations so they saw the same sky at the same time. Then, without moving the aim of their cameras, photographs took pictures whenever the aurora came into their field of view. To process their results, they laid one film over the other and moved one until the star pictures from one image matched as best they could the stars on the second film. Since the stars are far away and the aurora closer, they saw two images of aurora in the result. By measuring the displacement of the aurora on one film from the aurora on the other, they could mathematically calculate the altitude of the bottom and top of individual rays or bands of aurora. To do this, they had to precisely measure and know the azimuth and elevation angle of the point in the sky each camera was aimed at. Our goal was to make near-real-time height measurements of the base of fast-moving energetic aurora at television speeds of 30 measurements per second.

Using the television transmitter receiver system Herb provided, we were successful within moments of setting up the equipment. Our enemy was the midnight sun: each night in April the sun set later and went less far below the northern horizon. After only a few nights of successful operation, sunlight put an end to our measurements in late April 1976.

Hans, Neil, and Tom quickly analyzed several nights of observational data, wrote a research paper, and put my name down as lead author. *Geophysical Research Letters* had recently been created to get peer-reviewed papers into publication quickly, and the paper about our April observations was published in July.

The data showed energetic rays of aurora punching quickly down to altitudes well below 60 miles. We had expected these results, but now we had the hard data to prove how and when it occurred. As we closed down our observation program, I told Neil, Hans, and Tom that Herb had had some misgivings about giving me the television camera transmitter system. When they asked why, I told them it had last been used to transmit video images that operators had used to guide an unmanned air sampling drone aircraft into the radioactive cloud of a French nuclear weapon test in the South Pacific. I quickly added that Herb had assured me they had decontaminated the transmitter before sending it to us.

Launches in 1976–1977

By 1976, a resident staff at Poker Flat of approximately eight employees was supplemented during major launch operations by personnel from the various user organizations. At times, more than 50 people were involved in the launch operations. This highly flexible staffing scheme permitted rather complex operations at low overall cost.

During 1975, 198 Loki and 14 Arcas rockets had been launched for meteorological soundings in the altitude range 18–62 miles, and 12 larger rockets were flown to higher altitude for investigation of auroras and related phenomena. These rockets carried scientific payloads assembled by DNA, NASA Goddard Space Flight Center, AFGL, the US Army's White Sands Missile Range, Stanford Research Institute, Space Data Corporation, the National Oceanic and Atmospheric Administration, Aerospace Corporation, University of Arizona, University of Wisconsin, Utah State University, and the University of Alaska.

Meteorological Rocket Launches

Since June 1972, Poker Flat had launched small rockets three times a week for routine observation of temperatures and winds to 40 miles altitude. These observations were part of the ongoing program of the International Meteorological Rocket Network.

In addition to the routine meteorological rocket flights, several specialized projects required more frequent flights. Daily flights during January and June were performed in a project called Scale of Motion, intended to examine variations in the polar vortex that created the unusually cold 1976–1977 winter in the continental United States. Another project, in June 1977, involved the launch of 15 meteorological rockets to study stratospheric ozone and ion mobility in conjunction with a three-million-cubic-foot balloon flown from nearby Eielson Air Force Base. An operation in September 1976 required the launch of 30 meteorological rockets in association with balloon-borne cosmic ray and X-ray detectors flown by groups from the University of Denver, Utah State University, and University of Washington as well as larger rockets to measure incoming fluxes of auroral primaries.

A Loki-Dart sounding rocket launch from Poker Flat in November 1977. The solid propellant Loki-Dart sounding rockets were designed to measure temperature and wind velocity up to a height of 40 miles

High-Altitude Sounding Rocket Launches

The Meteorological Rocket Network team was actively involved in the first mission of the fall 1976 launch season. The project, Aurorozone I, was a comprehensive study exploring the possibility that X-rays produced by aurora might generate measurable changes in the density of stratospheric ozone. This September mission, led by Richard "Dick" Goldberg of NASA Goddard Space Flight Center, required that 13 Arcas and 17 other meteorological rockets be launched in a 9-day period. These Arcas were followed by three Nike-Tomahawks and simultaneous observations from balloons and by the Chatanika incoherent scatter radar operated by Stanford Research Institute adjacent to Poker Flat. Although the third Nike-Tomahawk blew up, it was ascertained that the premise was correct: they measured depletion of ozone density of up to 10% above 25 miles within hours of the onset of an auroral X-ray event. They also found that it took several days for the natural ozone density levels to be reestablished once depletion occurred.

November 1976 was an extremely busy month for Poker Flat launches. We did a particularly difficult launch series for Edward Fremouw of Stanford Research Institute, in which timing of the launches with the overflights was crucial. The experiment was designed to learn more about scintillations in radio signals passing through the auroral ionosphere, with a view toward improved satellite-to-ground communications. Precisely timed launches were required to coordinate with overflights of a satellite also instrumented by SRI. Two of four scheduled Honest John-Hydac rockets were launched to measure the ionospheric electron density with extremely high spatial resolution (3–9 feet) in a region being traversed by radio waves emitted by a passing satellite and being received at a special ground station installed at Poker Flat.

Also that month, another vehicle test involved the largest payload yet handled at Poker Flat: 2250 pounds. An 18,000-pound Talos-Castor sounding rocket system was used to carry a high-power electron accelerator assembled by Utah State University and a television system provided by the University of Alaska. The electron guns performed admirably, but because the system became unstable, camera data was limited and the recovery system failed.

Finally, on November 15, 1976, a test flight of a three-stage Talos-Sergeant-Hydac rocket system was conducted. The successful test provided the Geophysical Institute the opportunity to perform an experiment involving the release of water in the ionosphere for the purpose of modifying the electron density. We wanted to look into a phenomena first observed during the Skylab liftoff, during which widespread radio blackouts occurred, apparently as a result of the consumption of ionospheric ions by the water and hydrogen produced by the fuel dump of Skylab's enormous motors. By dumping a large amount of water into the ionosphere, we hoped to find out if that did indeed modify the electron density. Because the launch had to be made under less-than-ideal conditions, scientific results were marginal.

Immediately after the 2250-pound Talos-Castor launch, I perpetrated a hoax on Ed McKenna, the AFGL rocket payload/vehicle system mission chief. I had received a number of reports that Talos boosters remained surprisingly whole upon impact, with undamaged fins lying nearby. One summer day I set out to look at some of these sites myself. Most were across the river and on the hillside that faced the launchers at Poker Flat. I found a number of booster impact sites, including a few Talos booster impact sites. Often the Talos bodies and fins looked like they could easily be refurbished for reflight. Despite its weight and awkwardness, I hauled a fin back to the range. Rocket exhaust had scorched some nearby paint, but overall it looked nearly new. We eventually took it out of our lunchroom and propped it up against the side of the building, where it remained for several months.

That winter, teams from the AFGL and their contractors arrived at Poker Flat to launch one of their EXCEDE electron gun experiments. The rocket vehicle for this mission was a Talos-boosted Castor. As they prepared the Talos, we chanced to overhear that the Talos had a nasty habit of losing its fins during burn. I got the idea that we could put the fin I had brought back the previous summer on the ground near the launchpad right after launch. I asked Joel Lindsey and Steve Bonham to dig that Talos fin out of the snow next to the building and bring it inside to thaw out. Then we developed our plan.

First, we decided that we would attempt this stunt only if the mission were a success. Second, Joel would carry the fin around in his truck and only place it on the pad after the launch, when he and other Poker Flat crew nominally checked for damage. Third, he would call back from the launchpad and ask Ed to come down as he had found something he did not want to discuss on the intercom, or something like that.

We had our usual problems with the complex EXCEDE payload, and it was several nights later when everything came together and we launched. The Talos-Castor veered a bit away from the azimuth we wanted it to fly, but basically it flew fine. The EXCEDE payload functioned flawlessly, producing visible-to-the-naked-eye spots of man-made aurora throughout its few-minute flight. We waited in vain for the radio signals that indicated that the parachute recovery system had deployed properly. The prime telemetry signal quit abruptly at a time appropriate to a hard impact.

Everyone was excited with the success of the mission, and I gave Joel the nod to go ahead and leave the recovered fin on the launchpad. A few moments later I heard him on the range intercom calling for Ed. Ed broke away from his team of engineers, who were pulling paper records of telemetry signals looking for problems. After hearing Joel's message, he grabbed his heavy coat and called for a couple of his people to go with him to the launchpad. Soon thereafter the bruised fin lay on a table in the blockhouse with a number of people lamenting over what it might mean. The momentum of the situation kept building until Ed announced that he was going to have to call his boss Neil Stark at AFGL and let him know about this fin. My first attempts to intervene were unsuccessful, but before Ed completed dialing the number, I managed to convey that the rust on the fin should convince him that this was a hoax, and he put the telephone back on its cradle.

It was a good hoax, but in retrospect I have always had some misgivings about it. Emotions ran so high in the situation that I embarrassed Ed and others professionally, which I had not intended to do. No one ever really complained about the episode, and fortunately it never got outside of our little community of rocketeers. It remains one of my favorite yarns, told with care.

In January 1977, a group led by David Evans of NOAA in Boulder, Colorado, conducted one of the first successful scientific flights of the new two-stage Terrier-Malemute rocket system. The Terrier was a surplused military rocket, but the Malemute was built for NASA by Thiokol in response to the need for a second-stage motor that could reach higher altitude, in the 300- to 375-mile range. The rocket system carried an extremely complex payload instrumented to measure auroral particles. Four days later, a malfunction of the second stage of another Terrier-Malemute rocket system, under the supervision of John Lynch and Frank S. Scherb of the University of Wisconsin, prevented intended measurements of electrons and positive ions at high altitude in a proton-generated aurora.

In February we set a new altitude record for Nike-Tomahawk rockets. Due to a combination of a very light payload and an increased inter-stage coast time, this rocket climbed to 290 miles, almost double the usual expected apogee for this type of rocket. While this was partly a test to see if such high altitude could be achieved, the primary purpose was to perform shaped-charge barium releases. This was a University of Alaska project under the direction of Neil Davis.

The last launch of a major rocket during the 1976–1977 season was that of an Honest John-Javelin in June, partly as a proof test of a newly installed twin boom launcher and partly to verify performance of a rocket system modified by Space Data Corporation under Edward Allen's direction.

Recovering a 31-Inch-Diameter Payload

As Edward Allen of Space Data Corporation and I flew back to Poker Flat in the US Air Force helicopter, he leaned toward me and drew an impressively simple drawing of an idea he had come up with. I marveled at how he could make a clear and clean sketch as the helicopter shook and bounced along. I had no idea what he had devised because the noise level was so high that we could not talk and understand one another. What he had drawn looked like a cylinder with a ring around one end.

Behind us on the floor of the HH-3 helicopter were the recovered remnants of yet another failed 31-inch-diameter rocket payload. It was the result of the parachute being deployed down instead of up and behind the payload during the recovery sequence. We had found the drogue and main chute snarled in the debris.

Poker Flat Coffee Breaks

During a rocket campaign at Poker Flat, one visiting engineer—an athletic, good-natured professional from a major aerospace company—made a strong first impression. He was tall, wore glasses, carried himself with quiet confidence, and was well-liked by the local team for his easygoing manner and technical competence.

However, after his team's launch had taken place, several members of the Poker Flat crew overheard him making dismissive remarks about the LGBT community during coffee breaks in the Poker Inn kitchen. In response, a few of the crew decided to challenge his assumptions—not through confrontation but by staging a bit of unexpected theater.

Two of the male team members, known for their quick wit and camaraderie, began playfully presenting themselves as an affectionate couple. Their performance included light touches, cheek kisses, and lovingly exaggerated remarks. The rest of the crew went along with the act, treating it as completely ordinary and even adding friendly comments about what great family men the pair were.

The visitor's discomfort became apparent. He often got up to refill his coffee or left the room altogether. The intent wasn't to shame or humiliate, but to quietly remind him that Poker Flat was a place where respect for all people was the baseline—and that bigotry, however casually expressed, would not go unnoticed.

There was one difference between these 31-inch diameter payloads and the many others we had successfully recovered at Poker Flat. The others had a length-to-diameter ratio of about 7–1, whereas the 31-inch payloads had a length-to-diameter ratio more like 3 or 4–1. In developing successful sounding rocket recovery systems, an overall length-to-diameter ratio of at least 6 or 7–1 had been found necessary. A falling leaf or a long pencil will develop air resistance that causes it to tend to lie on its side as it falls, while a ball or a marble will not. All of the three or four 31-inch payloads with recovery systems that we had launched to date behaved somewhat like cannonballs, falling with no preferred orientation. In one particularly spectacular crash in the Brooks Range, the parachute had obviously deployed, only to be wrapped around the payload before landing.

The pilot lifted the nose of the helicopter slightly as we slowed and descended to the landing area at Poker Flat. Once we were on the ground, willing hands helped us load the debris into a nearby pickup. We rode in the back of the pickup to the Payload Assembly Building and then offloaded the debris for the night. By the next day, all of the debris would be packed up and shipped back to where it had come from.

Later, in the combination break and lunch room, Ed took his sketch out of his pocket and described his idea. To overcome the cannonball-like tumbling of the rocket payload during recovery, he proposed a slightly bigger diameter ring around one end of the payload. Ed's idea was to install a ring slightly bigger than 31 inches to one end of the existing 31-inch payload. He was confident that this would orient the payload during reentry so that the parachute would deploy properly. He wanted to build something like it back in Phoenix and launch it at Poker Flat as a kind of proof of concept. Ed proposed that we test this idea by flying a test object aboard a bare Nike rocket.

Officially, the launch fee at Poker Flat was set by the user agencies at our yearly meeting. The prime user agency was the same agency sponsoring these 31-inch payloads, Defense Nuclear Agency, and Ed and I were sure DNA program manager Herb Mitchell would approve. We told Herb we would use one of his Nike rockets we had in storage that no one wanted to put an expensive payload on because their storage history before coming to Poker Flat was missing. We had every reason to believe they were okay. It was the kind of can-do fun the space program was known for.

Over the next day, we talked it through. Ed would fly back to Phoenix and have the cylinder with the ring on it and a simple nose cone fabricated in Space Data's shops. They would be stacked atop a single-stage Nike rocket, and we would launch it east out of Poker Flat to land a mile and a half away on our land. Ed and I would stand near the impact point of the rocket to observe whether the cylinder with ring came in oriented with the ring acting as a drag.

We pulled a DNA Nike out of its box and installed and set the fin angles for as little spin as feasible. I kept expecting a carefully machined and checkerboard painted test body and nose cone to arrive for our test, but when the test body and nose cone finally arrived, the body was a ragged hunk of pipe the same diameter as a Nike with a crudely machined ring about an inch and a half bigger in diameter bolted within an inch or so of its aft end. The nose cone looked even worse. All of the slip joints were so loose that I was concerned it would fall apart prematurely. When I asked Ed about the lack of a checkerboard paint pattern to assess roll, a roll of masking tape and a can of black spray paint immediately appeared. A stripe was laid out and painted on the side of the test body. More spray paint was applied to ensure observers could tell which end of the body can had the ring on it. I suppressed any misgivings and wholeheartedly joined in the last-minute planning.

We loaded the Nike rocket on pad 2 in a horizontal position and slipped the cylinder over the head cap area. The slip fit between the cylinder and the Nike was good and loose and yet long enough that it did not fall off. Ed loaded the cylinder so the ring end would be just beneath the nose cone. He was sure the nose cone and cylinder would drag separate from the Nike near apogee, but he wanted the drag ring to turn the cylinder around so it came in with the ring at the rear.

It was a cool fall day, not cold, so we did not bag or try to keep the Nike warm as we often did for winter launches. It was late afternoon, and the sun was low on the southwestern horizon. We elevated the launcher to 85°, aiming the rocket about a mile and a half away on land that was owned by Poker Flat. We were confident we could predict the impact point. We chose to launch almost due east so as not to cross the Steese Highway. Ed, Beau Battey, Joel Lindsey, and I stood about 300 feet from the predicted impact point and spread out so there were a hundred or so feet between us and then told Carroll Coe by radio that we were ready for him to launch the Nike.

The Nike did what it does: burn 800 pounds of solid fuel in 3.5 s and make a huge amount of noise, and then coast up to apogee and back down. But this time, without an upper stage, it topped out near 30,000 feet, not 8000 feet. I had not fully anticipated how high it would go and held my breath for a few moments when I saw it break out of the winter valley shadow into brilliant sunshine and continue upward. We could not see the Nike or its payload, only its trail of smoking propellant. Brief jokes about not hearing the bullet that hits you passed among us. It slid back down out of the sun, and a minute or 2 later, I could make out all three pieces: the nose cone, the cylinder, and the empty Nike booster rocket casing with fins. Everything had aerodynamically drag separated at altitude just as planned.

It fell with increasing speed before eventually hitting the ground at just under 500 miles per hour. It landed where we wanted it to, a good 300 or 400 feet away from where we stood. The Nike still weighed 800 pounds when it hit the ground. It dug a relatively small crater no more than 8 feet in diameter and 3 feet deep in the rocky soil. A puff of smoke rose now and then out of the steel case in the dirt crater. I still feel it was one of greatest and perhaps craziest things I have ever done. How many people have ever purposely stood so close to a potentially lethal event? I still grin remembering what it felt like.

The cylinder with ring looked about as good as it had before it was launched. We knew it was an extremely loose slip fit and easily convinced ourselves that air pressure on the ring had kept the recovery body pressed against the Nike throughout the flight. The nose cone came in last, seeming to alternate between tumbling and floating. Hitting the ground warped it out of round a bit.

As soon as we got back to Poker Flat, Ed called his home office in Phoenix to tell them that it worked.

Space Data elaborated on this concept to build a spring-loaded expansion ring near the aft end of a 31-inch diameter recovery section. It was about 45 inches long and painted yellow. Twenty or more pieces of cast metal that honestly looked like lamb chops were mounted in a slot near one end of the section. They were spring-loaded to pop out of the side and give it a bigger diameter on one end, similar to what the ring had done for the Nike cylinder. They were perhaps an inch thick. Because Space Data did not want them to exert a drag load during the launch phase, they were held in by a surrounding breakaway band that increased the diameter by only an eighth of an inch at most. When deployed they added roughly 8 inches to the diameter. The breakaway band would be released at about 50,000 feet on the way down. Inside the yellow cylinder were a conventional heat shield, drogue chute, and main chute recovery system.

The following summer we launched a successful full-up engineering test aboard a 31-inch Javelin-boosted Sergeant rocket. The lamb-chop orientation recovery system for the payload worked perfectly, and Space Data used this system successfully for all of the many payloads they launched for Defense Nuclear Agency.

A Record Heavy Payload Recovery

On October 19, 1979, we launched a Space Data Corporation Talos-Castor system for the EXCEDE II mission, weighing 8800 kg (19,400 pounds) at launch. It boosted a payload weight of 2586 kg (5701 pounds) to 79 miles altitude. This was the largest scientific payload and total vehicle weight ever launched from Poker Flat, and perhaps anywhere.

EXCEDE II was a DNA-funded AFGL experiment to create and fly electron guns that would irradiate and make sophisticated spectral measurements of prompted emissions in the atmosphere near the payload as it flew. The intense electron gun beam was a controlled nondestructive experiment that would simulate what would happen if a nuclear weapon was exploded at these altitudes in the Earth's atmosphere.

The payload separated from the rocket motors at 43 miles and despun at 47 miles, and the cover doors for the scientific instruments and electron guns were ejected for the scientific experiment to start. An attitude control system oriented the payload so that the scientific instruments looked up the local magnetic field while the electron guns irradiated the volume they were looking at. The guns started firing at 59 miles and continued firing through apogee at 79 miles until they were shut off again around 59 miles on the down leg of the rocket's flight. I stood outside after launch and remember seeing the electron gun pulses create visible searchlight glows in the atmosphere each time they fired.

At 50 miles on the down leg, the 3690 pounds of batteries and electron gun portions of the payload separated from the 2000 pounds of scientific instrumentation that was rigged for parachute recovery. The radar system at Poker Flat lost track of the radar beacon onboard the payload at 410 s after launch. The telemetry system at Poker reported loss of signal 437 s after launch. Data showed it landed approximately 125 miles from Poker Flat, in the White Mountains, about 30 miles southwest of Central, Alaska.

We quickly located it the next day in a fixed-wing aircraft. There was no snow yet in the area, and the payload was in a wide creek bottom, smashed to smithereens with no sign of a recovery parachute.

We had a memorandum of understanding with the US Air Force search-and-rescue group at Eielson Air Force Base to recover payloads once we had an eyes-on-the-ground location. When we found something and contacted them, they would call back and tell us when they could help do a recover. They used our recoveries as a crew-training mission for their helicopters and crews.

When the HH-3 helicopter for this recovery arrived at Poker Flat 2 days later, I asked the crew if Randi Wagner could come with us, and they said yes. It was a pleasant warm midafternoon when we landed to one side of the dry creekbed. We could see that the payload had landed on its side. About one-third of its 31-inch diameter had been flattened. The hundreds of thousands of dollars of scientific instruments had been spit out to one side during the crash and lay in pieces on the ground.

We could see that the drogue chute had deployed, but the main chute had not. Over the next hour, we picked up pieces and put them in large nylon mail sacks we kept for just this purpose, rolled the carcass into a cargo net, used the cargo winch in the HH-3 to pull it aboard, and returned to Poker Flat.

Overnight while pawing through the wreckage in our Payload Assembly Building, the folks from Space Data realized a couple of items that could give us clues as to why the recovery system had failed must still be out there. The crew from Eielson Air Force Base came back and flew us out to the area, but it was now covered with about 6 inches of fresh snow. We never went back to the site.

Spire Launch and Recovery

Ralph Haycock from Utah State University was a superb mechanical engineer who led the liquid helium support work for AFGL rockets launched from Poker Flat. Spire I was launched in 1977. We counted it down for 89 days before we finally launched it. We were on range for 16 h for each countdown—a range record. The last night, the launcher was elevated to about 45°, and the launch conditions finally looked great, except Ralph said he and his team needed to refill the payload with liquid helium to have the launch be successful.

Fording the Chatanika River

One June weekend day in the late 1970s, I decided to walk across the Chatanika River to look at the damage caused by first-stage Nike rocket motor casings when they landed. My path would be along the primary magnetic north launch azimuth for Poker Flat, which led up a small hill covered with spruce trees. I planned to walk across the Chatanika River on the gravel bar that lay less than a foot below the surface of the water.

The river was only 75 feet wide, but the gravel bar ran about a hundred yards on a diagonal from the south shore to the north shore. I had crossed the Chatanika River at this point several times before in four-wheel-drive pickups in the summer and in tracked vehicles in the winter when the river had frozen over.

Before crossing, I took off my hiking boots and put them in my pack and put on tennis shoes. The water was not running hard, but it was cold. The water in Interior Alaska rivers is frozen seven to eight months of the year, and when it melts it usually runs over frozen ground before getting into the rivers. I doubt the water in the Chatanika River was much more than 32 °F. A quarter of the way along the gravel bar, my feet were painfully cold. The morning was cool but the sun was warm, so I continued on rather than turn back. Halfway across, my lower legs and feet ached, but I knew it would hurt just as much to go back as to keep going.

By three-quarters of the way across, I had trouble moving my cold-numbed lower legs and feet, and I thought the pain was too much. I decided to step off the gravel bar and leap for some branches that hung low over the river only a foot or so away and pull myself over to the north bank. This turned out to be a bad idea. The river bottom was well below the reach of my feet, and I and my pack went almost completely underwater. I hung onto those branches for dear life. The current caught and swung me in a downstream arc to the north bank and, fueled by an adrenaline rush, I pulled myself up the bank. I gasped for breath and thought to myself, "I don't think I have taken a breath since my chest hit the water." Only then did it sink in that I had done something really dumb.

I shivered and shook and stripped off my wet clothes and put on the warm dry ones I carried in my backpack. I thought about starting a fire but decided walking would warm me up faster. I left my wet clothes draped over some bushes to dry. I would recover them when I recrossed the river later in the day.

I spent the morning wandering from rocket crater to rocket crater up the hillside. Each crater consisted of disturbed ground about 8 to 10 feet in diameter and 2 to 3 feet deep, with an 18-inch-diameter crumpled iron rocket motor casing standing in it. The rocket casings had originally been about 12 feet long but were compressed accordion fashion after hitting the frozen ground. The heavy iron nozzle at the aft end of the rocket usually stood clear of the ground. The four fins that had been mounted around the nozzle had usually sheared off and lay on the ground beside the casings. I knew the ground had been frozen when the empty 800-pound rocket casings had impacted at about 400 miles per hour and was surprised at how little damage to the ground had occurred.

By noon I had made it to the top of the hill. There I discovered the 30-foot-wide shallow crater that had resulted from the impact of a live second-stage Apache rocket. It had blown up on impact there years before. I recovered some unburned propellant that looked like sponge rubber with aluminum flakes in it and put it into my backpack to look at more closely later.

After lunch I walked from the ridgeline downhill to the Davidson Ditch that had been built to bring water from 60-mile on the Steese Highway to the gold dredges nearer Fairbanks.

Poker Flat is at mile 30 on the Steese Highway, and the Davidson Ditch was about 200 feet higher in elevation than the Chatanika River at this point. I walked along the Davidson Ditch to a siphon that led from the north side of the Chatanika River valley to the hills on the south side.

I also explored a small abandoned house near the top of the siphon. The people who lived in this house in the summertime maintained a few miles of the ditch. Along the southern edge of a portion of the Davidson Ditch was a broad field of waist-high grass on a gently rolling downslope. The sun was warm, and the grass moved gracefully in the wind as I walked through it.

Then I recalled that one of the biggest grizzly bears ever killed in Interior Alaska was taken not many miles from where I was now walking. I stopped and took my.357-magnum pistol out of my backpack, loaded it, and strapped it around my waist so it would be easier to get to if I needed it.

In late afternoon I stood again on the north bank of the Chatanika River. I put on the clothes and tennis shoes that I had left on the bushes by the riverbank after my morning plunge. They were still damp.

I quickly walked the 100 yards along the gravel bar to the south bank and dry land again. I was able to move faster now as I was walking downstream. By the time I reached the south bank, my feet ached with cold, but I had no inclination to plunge into the water again to get to the shore faster as I had in the morning.

Liquid helium is treacherous stuff. A liquid helium dewar was about 5 feet in diameter and 6.5 feet high. A few hundred liters of 4-degree liquid helium were in an inner dewar, surrounded by another dewar filled with liquid nitrogen.

Ralph asked if they could use our high bucket truck to refill the payload with liquid helium. I talked to our launch officer Carroll Coe, and he said he would disarm the rockets. We trusted Ralph and his crew, so we let him reservice the payload. Later, Ralph mentioned that he and his team would never have been able to service an elevated payload at any rocket range in the world like they did at Poker Flat. It would have violated range safety rules.

The rocket launch for DNA and AFGL was near perfect, and the data acquired in aurora while it flew was stored safely on NASA telemetry tape recorders. As the payload passed through 70,000 feet altitude on the down leg, 240 miles northeast of Poker Flat, we could no longer interrogate its radar beacon from the tracking radar at Poker Flat.

Over the next several wintry days, Poker Flat's Steve Bonham flew over the predicted impact area in a fixed-wing aircraft without hearing the radio recovery beacon that should have been activated at 20,000 feet when the parachute was deployed. It would normally transmit a recovery radio signal for at least 3 days, even at 50° below zero. Because he could hear nothing and because he saw nothing—no parachute, no churned-up snow or scarred landscape from a crash—we presumed the recovery system had failed. We terminated the search, planning to go back again in the summer when the snow had melted.

That summer Steve spotted the Spire payload and nose cone far north of where we thought they might have been. The bright aluminum debris lay in a clump, sparsely covered with light brush, in a wide, nearly flat alpine valley of the Brooks Range east of the Sheenjek River. Steve marked the location on a map. After he returned, we asked the Eielson AFB search-and-rescue group to help us recover it.

Early morning one fine summer day, the helicopter landed at Poker Flat. The HH-3 crew consisted of a pilot, copilot, and crew chief. This looked like a great opportunity for me to go along on a recovery mission and take pictures. Beau Battey worked with the sergeant scheduling operations for one of the Alaska Air Command's HH-3 helicopters as a search-and-rescue training event. Beau had been a combat parachutist in the military, and I took his warnings about how to prepare for this trip seriously. In particular, he had cautioned me about the Brooks Range bluebottle flies, which were not only huge but vicious in clinging to open skin.

Carroll initially surprised me by saying he did not want to participate. Early in his air force career, he had been part of a B-36 bomber crew, and he told me frankly that while he now trusted commercial flight operators to properly maintain their aircraft, he knew from his own experience that maintenance of air force aircraft was not as good as for commercial flights. However, we were really short of people to help recover this payload, and Carroll reluctantly agreed to come along.

We flew uneventfully about an hour north of Fort Yukon. As we approached the location, we could see that the payload lay in an area that was reasonably flat and covered with large, well-washed river gravel. This would make a stable landing area for the HH-3. The HH-3 is a relatively large helicopter, and as we hovered in our descent, the wind under its main rotors swirled and twisted the knee-high brush. Beau had indicated to the pilot that we would be on the ground about an hour. After landing, we sat inside until the rotors stopped turning and the main engines were shut down. As we climbed out and offloaded some heavy canvas sacks for debris and some tools, I looked up to see the pilot and co-pilot had pulled some sort of stocking tube sacks down over their heads and were tucking them beneath their flight coveralls. They then put their sunglasses back on and continued to put on gloves. The covering I had on in response to Beau's suggestion was flimsy in comparison.

It was a bright, sunny day. The temperature was warm, not hot, and a light breeze made it really pleasant. Steve had been out like this several times before, and I asked him why the pilots and crew chief had covered their heads, because right now I neither saw nor heard any insects. He grinned and said, "You will." He said the downward blast of air under the helicopter blades as we were landing had caused all the bugs nearby to grab hold of leaves and other brush and hang on for dear life, but in a few minutes we would see more of them than we wanted.

The rocket payload lay broken to pieces squarely in the center of the recovery parachute. Most of the parts were close together, with only a few thrown more than 3 or 4 feet away. The payload had come in with the expensive infrared telescopes pointed downward. They had crumpled accordion-style at the impact. The rear telemetry deck, attitude control system deck, and recovery decks were in surprisingly good shape. The recovery heat shield was gone, the small drogue parachute that pulled out the main parachute was deployed, and the main parachute bag was out of its canister, but the razor-sharp circular cutters attached to the small parachute had not cut the cords that held the main parachute in its bag. Clearly everything had almost worked. Only a couple of seconds or a thousand feet of altitude had stood between success and failure of this recovery.

I took a few documentary photographs, and then we started gathering loose pieces into bags and putting them onboard the HH-3. I overheated in my bug covering and took it off, as did Beau. The HH-3 crew including the pilot and co-pilot helped us haul the debris back to the HH-3. One major chunk was too heavy to lift or drag up the ramp at the back of the HH-3, so Steve, Carroll, and I worked it into a cargo net and rigged cabling for the helicopter to carry it as a sling load back to Poker Flat.

The bugs slowly came out but were not very bothersome. The noise of their buzzing had grown steadily all the time we were on the ground. But with about 15 min to go, I suddenly felt something large grip the back of my left wrist and plunge into my skin. I looked down and saw the biggest fly I had ever seen outside of an insect collection in a museum. Its colors were iridescent blues and blacks. It was beautiful in a sort of sinister-looking way. It did not shake off, nor did it budge when I tried to scare it away by waving my other hand close to it. I had to physically pull it off. It released its hold on me reluctantly but left no visible scar. I rolled down my sleeves, reinstalled my head net, put my gloves back on, and kept working.

Over the next 10 min before we finished, several more landed on my clothing and one or two more on exposed skin. A few more bug monsters effectively gripped right through my shirt into my tender skin. In each case I had to pull them off to dislodge them. We loaded our debris back into the HH-3 and prepared for takeoff. It felt a hellish eternity fighting bugs inside the solar-heated fuselage before the pre-flight checklists were completed and we were airborne out of there. The crew chief left the sliding side door open, and the bugs somewhat miraculously left us.

We carried the sling load about a hundred miles from the impact site back to Fort Yukon, where we stopped for more fuel. While there, we commandeered a local forklift to help us maneuver the payload onto the rear ramp and used the winch inside the HH-3 to pull it inside. After we landed at Poker Flat, we were able to offload the payload by rolling it down the ramp.

Blinded by the Light

In February 1978, the launch of a never-before-flown combination of three staged rocket motors—a Talos booster, a Sergeant sustainer, and a Hydac upper stage—had gone smoothly. At 93 miles, the payload separated from the upper stage and exposed a downward-looking television camera. Radiotelemetry from the payload told us everything was working well. We watched signals telling us that despin weights had been released. Two weights, each about half the size of a human fist, one on either side of the payload, had been released and were now unwrapping cables that connected them to the payload. Like an ice skater in a fast spin who stretches their arms out, the payload slowed from two revolutions per second to a tenth of a revolution per second.

Radio telemetry told us the experimental attitude control system onboard was now aligning the payload into a vertical position as it flew higher. It had been a nighttime launch, and the backward-looking low-light television camera onboard transmitted the slowly rotating images of the lights of Fairbanks and Eielson Air Force Base.

On its way to a predicted 310-mile peak altitude, the television images went pure white at 215 miles. We broke our sudden blockhouse silence by asking telemetry and radar if they were still tracking. They both replied yes, all the radio signals, including those from the television transmitter, were solid.

In the blockhouse we watched the white screen of the television monitor for any sign that the television camera, which had apparently failed, might recover. Bill McKecknie of Defense Nuclear Agency, Ed Allen of Space Data Corporation, myself, and several others were puzzled by the nature of the failure of the television camera but told one another that the real objectives of the launch, the test of the rocket motor systems and attitude control system, were successes.

Flying Backward in a Helicopter

I'm seated beside the pilot in a small two-seater helicopter, churning slowly forward just above the treetops as we search the hillside for debris from a Strypi rocket that had exploded the night before. It's my first time flying with this particular pilot in this particular aircraft. As we crest the top of a hill, I expect him to increase power, gain altitude, and arc down to the base to begin our next search path up the slope.

But instead, to my surprise, we begin descending the hill tail first—slowly, steadily, and still at treetop level. I don't think too much of it at the time and continue scanning the foliage for signs of wreckage.

After a couple of hours, we break off the search and return to Poker Flat for refueling. Curious, I ask the pilot about the unusual reverse descent. He replies casually, "Something I learned in Vietnam," and leaves it at that.

A new set of eyes takes my seat for the next flight, following a similar search pattern alongside the area I had just covered. None of us turn up any debris that day, and we regroup at the Poker Inn break room to compare notes and plan tomorrow's search.

Beau Battey and Steve Bonham—both far more experienced than I flying in helicopters over busted-up rocket sites—comment on the remarkable skill of the pilot I'd flown with. They're especially impressed by his ability to back a helicopter down a hill with such precise control, skimming the treetops all the way.

I mention that the pilot told me he'd learned that technique in Vietnam and hadn't thought much of it. Others in the room nod, noting that Vietnam veterans who flew helicopters were often extraordinarily skilled. Many had developed techniques—like backing down a hillside—that were neither taught nor encouraged, but born of necessity and survival.

There were several Vietnam-era helicopter pilots around Fairbanks in those days, but this would be my one and only experience flying backward in a helicopter.

After tens of seconds had passed by, the image of the lights of Fairbanks and Eielson Air Force Base from the television camera returned.

As I recall, I was the first to figure out what had happened: the payload had flown up into pure or significantly scattered sunlight. I told everyone that the television camera had been overwhelmed with light, thus the white image. I asked what kind of shielding had been placed about the television lens along the optical axis of the lens and learned that it had none.

Earth's orbit is roughly a circular path around the equator of the sun, and Earth spins on its axis once per day. The deepest dark of night is when the sun appears to be on the opposite side of the Earth from where you are. But if you ask someone who lives in Fairbanks, at roughly 65° north latitude, how dark is it at midnight, almost everyone will say it depends on what season it is. You can no longer see the stars, or aurora, from roughly the end of April until the first of August because of sunlight scattered by the Earth's atmosphere, which is a thick blanket of gases that surround the Earth. During summer solstice, roughly June 21, the sun never gets more than 4° below the northern horizon at midnight, and during winter solstice, roughly December 21, the sun never gets more than 4° above the southern horizon at noon.

Now picture the reality of a 864,000-mile-diameter sun, 93 million miles away from Earth, and the configuration they are in for it to be midnight on Earth. Now think about what midnight must be like if you lived at either the North Pole or South Pole. There, because the spin axis of the Earth is inclined at 23.5°, you would experience weeks each year where the sun never sets and weeks where the sun never rises.

If you are at the North Pole on February 28, the sun never gets more than 8.5° below your physical midnight horizon. Scientists call this your solar depression angle. How high above your head is sunlight streaming through the Earth's atmosphere across the top of Earth when the solar depression angle is 8.5°? You can do the math, but a trick scientists use is to multiply the solar depression angle times itself, and then tell you that the resulting number is how many miles over your head the sun is shining.

Now that I have laid this groundwork, I am going to tell you that at 11:44 Coordinated Universal Time on February 28, 1978, at Poker Flat Research Range at 65° north, the solar depression angle was such that Earth's atmosphere scattered enough light into the unshielded lens of the rocket-borne television camera to cause it to show only a white blank screen on our monitors at Poker Flat.

Disarming Wescott Barium

Late one night in February 1978, our payload engineer Lou Baim quietly told me we had a problem. The safe-arm system for the 20 pounds of high-explosive material in the nose cone of the rocket on pad 3 was stuck in the armed position.

Lou told me that he was unable to disarm it by himself. I told him at least three people needed to be involved, with one doing the work and another alongside them, communicating what they were doing to a third person in the blockhouse. The person in the blockhouse would keep written notes, just in case something went wrong and the two doing the disarming were injured. I was the most familiar with what Lou had built, so I went down to the launchpad to take the payload off the rocket and later disarm it.

This payload had been built to carry out an experiment Geophysical Institute scientist Gene Wescott had devised. The very first launches from Poker Flat in early 1969 had involved rocket-borne releases of barium vapor directly into aurora. Neil Davis, Gene, and other scientists at the Geophysical Institute wanted to use barium release technology to study particular events in the aurora. The first experiments were not very successful. The scientists wanted to align a shaped-charge release of barium vapor along the Earth's magnetic field lines in aurora, and this cost more than the funding they were initially able to secure from the National Science Foundation. But they had acquired valuable knowledge about how to build barium payloads.

The Geophysical Institute scientists worked a deal to secure funding from NASA wherein they led the science and Los Alamos National Laboratory (LANL) scientists and engineers provided rocket motor systems and dunce-cap-shaped barium liners for shaped-charge explosives. We had a number of spectacular successes in this arrangement, but by 1978 LANL had lost the internal funding needed to participate in the joint project and dropped out.

Neil and Gene wrote successful proposals to NASA to build barium shaped-charge payloads that would be launched onboard NASA Nike-Tomahawk rockets. The rocket payloads were little more than a shaped charge with a barium liner, a nose cone that was automatically ejected 90 s into flight to expose the barium canister to the atmosphere, a timer to set off the explosives, and a magnetometer that would complete the onboard firing circuit when the payload spun into alignment with the Earth's magnetic field. The payloads typically weighed less than 60 pounds.

The partnership between NASA and the Geophysical Institute created some tensions when it came to safety. NASA knew all about rockets, ejectable nose cones, timers, etc. but did not want high explosives in their test and integration facilities for sounding rocket payloads at Wallops Flight Facility in Virginia.

Technically, the explosives part of the payload was Lou's responsibility. The NASA Nike-Tomahawk rocket was the responsibility of Jay Brown, the NASA mission head, and Gene was the scientist for the experiment aboard this rocket. Lou had designed and submitted the safe-arm assembly design to NASA Wallops. They approved it, and he constructed it and the holder for the shaped-charge explosive with its barium liner in Alaska. Gene found a company in Denver, Colorado, to cast the shaped charge, and DuWayne Bostow machined stock barium metal to match the shape of the explosive in a special building at Poker Flat. The shaped-charge assembly and NASA's payload and rocket assembly met and were mated for the first time at Poker Flat, about 2 weeks before each launch.

When I showed interest in the arming system one day, Lou proudly showed me how simple it was. Lou had built the safe-arm unit by machining a chamber and a passage in a small block of aluminum and installing a cover plate. On the cover plate, he attached the electrically fired igniter to an arm that was moved by a small electrical motor so that it would direct the igniter's flame either through a passage into the explosives or into the chamber where it could not reach the explosives. In the armed position, the igniter was aligned to blast right into the base of the explosive. In the safe position, the arm was rotated out of alignment with the shaped charge and into alignment with a small chamber that would capture and cool the igniter gases to keep them from accidentally setting off the explosive.

Before coming to see me with the problem, Lou and the crew had disconnected all electrical wiring between the blockhouse and pad 3 where the rocket and payload hung. On the launchpad, Lou and I told the other crew there to go back to the blockhouse and leave us by ourselves as we worked on the payload. First we put in place a canvas strap attached to the launcher to keep the payload in place as we removed the bolts that coupled the payload to the upper rocket motor stage. The payload was 9 inches in diameter and about 40 inches long and weighed about 50 pounds. After we were sure it could not accidentally fall off, we did little more than disconnect the umbilical wires that connected the payload to the launcher for system checkout, take out the bolts that held the payload to the second-stage rocket motor, and pull the payload forward and away. It was not heavy. It was easy for the two of us to handle. Lou and I put the payload in the back seat of a car and drove a hundred yards to Rocket Assembly Building B. Once inside B-RAB with the payload, we established communication with the blockhouse again and pulled out the safe-arm assembly. After we had done that, the explosives could no longer be accidentally ignited, to the relief of everyone on range.

The swing arm that had worked perfectly many times before was jammed. It and all the pieces involved were fabricated from soft aluminum. When we dismantled the parts, we found the soft aluminum of the arm in the chamber had scratch marks and slivers of peeled aluminum in it, so it was jammed open and would not move. We decided that if we could find, cut, and place a thin metal spacer in between the cover plate and the chamber, we could have everything back together, working properly, and back on pad 3 ready for launch. We needed a sheet of ultrathin aluminum to space the machined pieces just a tiny bit farther apart. We looked in a nearby trashcan and remarked almost simultaneously that we had found the perfect answer. Lou plucked out an aluminum drink can and cut it open with his pocketknife. It was big enough and thin enough to do what we wanted. He cut it with a pair of scissors to the exact shape needed, and we reassembled the safe-arm assembly and operated it. It worked perfectly.

Less than 2 h elapsed between when Lou first told me he had a problem and when we had the payload back on the launcher, elevated, and ready to launch into aurora.

Lou and I started to tell Jay, the NASA mission head, how we had fixed the problem, but he put his hands over his ears and with a grin said, "I don't want to know." He did not want to submit our fix to NASA for safety concurrence. If anyone above him asked, he planned to tell them that we had taken it apart and adjusted some clearance. He hoped they would not ask.

Death Threat

For a few weeks one summer, big thunderclouds formed over Poker Flat nearly every afternoon. Heavy rain fell from these clouds as they drifted generally south. One afternoon I received a telephone call from a man who said he knew we were creating these storms by seeding clouds with our meteorological balloons and rockets. He was angry and refused to believe anything I said. He called several days in a row, and then one morning he vowed that if it rained that day he was going to kill me.

I contacted the Alaska State Troopers, who said they would look into it. I also told the director of the Geophysical Institute and the campus police, neither of whom appeared to take it very seriously, which was a bit unnerving. I started carrying a loaded 0.357-magnum pistol under the seat of the truck I drove to Poker Flat each day.

The *Fairbanks Daily News-Miner* started printing letters to the editor that implied Poker Flat and the Bureau of Land Management were seeding clouds to create the unusual thunderstorms. At a public forum, members of the community and press asked me and Bill Robertson from the BLM what we knew about seeding clouds. Bill and I both said we were not currently seeding clouds.

Bill added that in the 1960s the BLM had in fact seeded clouds west of Fairbanks in an attempt to keep them from becoming so powerful. This was based on the idea that moisture-laden winds coming from the west toward Fairbanks were lofted by rising air created by solar heating of the Tanana uplands. This was considered to be the probable cause of the lightning associated with thunderstorms that was causing a number of forest fires. BLM seeded the clouds west of Fairbanks using balloons carrying cloud-seeding chemicals. They also dropped chemicals from airplanes, probably silver iodide, which has been used in other places and serves as a nucleating agent for water vapor to become water drops.

Bill said that the experiment did not work and that the letters to the editor were probably initiated by a disgruntled former employee or contractor who as a pilot had hauled the cloud-seeding materials to the remote sites.

I kept pressure on the Alaska State Troopers to defuse the death threat. A few weeks later, they told me that they had found and talked to the individual who had threatened me and that they did not think I would receive any more threats from him. They said they thought he was harmless, but they would not tell me who he was.

A few nights later, the conditions were right and we launched. The rocket flew high, and at the right time, the shaped charge was fired, creating the successful beginning of an aurora experiment. Only Lou and I knew that the ultrathin sheet of aluminum that "adjusted the clearance" came from a Budweiser beer can.

You Shoot 'Em, I'll Load 'Em

As we were discussing our workload on March 1, 1978, Carroll Coe said, "You shoot 'em Neal, I'll load 'em." If I thought we could meet the science objectives with a launch, Carroll would lead the crew to make it happen. The next day we did just this with a launch for Mike Kelley of Cornell University and T. Stockflet Jorgensen of the Danish Meteorological Institute. Their launch was a chemical release program involving photographic observing stations without communications and far from Poker Flat. By prior agreement, we would launch at a specific time each evening or morning, and each remote station would take several tens of minutes of pictures only if they saw the chemicals start at the given times.

Around 4 p.m., launch conditions for their NASA-sponsored launch were looking good for the agreed-upon launch time of 6:23 p.m. But we had problems at Poker Flat. First, only the pad 4 launcher was available, and it had been damaged a few days before during the launch of a Sergeant-Hydac-boosted mission. Second, although the Nike-Tomahawk rocket motors were built up, they were still in B-RAB. And third, the TMA canister for their payload needed to be brought up to a specific temperature to work properly. The TMA canister had an electrical band heater, and the flight requirements document stated two hours of heating would be needed.

I asked Carroll if we could load and launch the Nike-Tomahawk system on pad 4 without protective Styrofoam heat covers, since it was a relatively warm, sunny afternoon under clear skies. Carroll said yes.

I talked with Chuck Keirstead from Thiokol and told him the range could support a launch of his payload in 2 h. He responded that for his TMA canister to work properly, he had to plug in some electrical heaters, and it would be really marginal for him to get the TMA canister up to proper temperature.

I notified Mike Kelley that we were going to try to make the 6:23 p.m. launch time, and told Randi to count it down as if we were going to launch with no holds.

Chuck plugged in his TMA canister heaters in A-RAB, and Carroll and his launch crew moved a built-up Nike out of B-RAB and down to pad 4. The boom on pad 4 is roughly 12 feet above the ground, so Carroll and his crew used chain-falls on the I beams on the launcher and straps to lift the Nike into mating position with the launch rail, and then slid it back to the first motion switch stop. Carroll and his crew went back to B-RAB, brought down a built-up Tomahawk, and did the same.

At 6:10 p.m. Mike wondered out loud if we were going to be able to launch their rocket. I could only tell him I was pretty sure we were going to make the 6:23 p.m. target. We had helped Chuck load his payload of barium and TMA canisters to the launchpad, and by this time, it was mated to the rocket motor system. We jury-rigged a Herman Nelson heater hose so it directed heat onto the payload.

Thiokol's blockhouse ground support equipment continued to heat the TMA canister through the umbilical cables, and thermistors on the canister confirmed we were getting closer to an acceptable temperature for launch. The payload had only barium and TMA canisters activated by onboard timers. No other experiments were in the payload, and there was no telemetry or radar beacon.

I was on pad 4 with Carroll and our technician Joel Lindsey, and with minutes to go, Joel looked up at the pad 4 first motion switch and saw something was still not right. Joel and Carroll put an A-frame ladder under the first motion switch, so Joel could climb up and properly align the first motion switch rod. We laid the ladder down nearby.

Carroll used local controls at the launcher to elevate and rotate the launcher to approximate firing position. I used the pad 4 intercom to tell Randi in the blockhouse to send out roadblocks to block the Steese Highway in front of our launchers and to tell Mike we were going to launch on time. I remember hearing Randi announce we were at T-4 minutes and counting just before Carroll, Joel, and I jumped in our vehicle and sped back to the blockhouse.

When I got to my position beside Randi at the launch control desk, she said we were going with the FAA. Carroll used the remote launcher controls in the blockhouse to put in final launcher settings. Chuck declared the TMA canister temperature was go. Carroll checked his firing panels and said he was go. We launched on time into a beautiful star-filled evening sky. PF-NT-139 flew perfectly and dispensed its barium and TMA into a sunlit ionosphere above us. The photographer who had been sent to Arctic Village told me later it was a beautiful release. Mike was a very happy rocket scientist.

Pictures of barium releases like this are taken over a period of tens of minutes. The barium ionizes in sunlight, and the pictures show both the neutral and ion cloud. The motions of the ion cloud over time are driven by the same forces that create what you see in the aurora, and so are used to better understand the electrical and magnetic field structures present in aurora. TMA is dispensed as a vertical trail, and photographs of it over time give a sense of the lower ionosphere wind speeds and directions.

Ed Allen of Space Data Corporation, who had been involved with the Sergeant-Hydac launch that had damaged pad 4, was on range too. He came up on the range net and said, for everyone to hear, "Congratulations, you guys did it again."

I thanked Carroll for his efforts after the launch and thought to myself we had done it again: "Neal, you shoot 'em, I'll load 'em."

Burglary: Stoners, Mosquitoes, and Peanut Shells

We had just finished up a spring launch operation in 1978. It was a sunny afternoon, and people were lazing around on the front porch of the office building before heading home. I was living in Poker Inn at the time, and I went home to put on my running clothes and then ran up to the Loki launcher on hilltop and back down. As I was finishing my run, I noticed one of our gray Dodge pickups sitting in front of the Payload Assembly Building, blocking the front door. I was irritated that it was left there. I opened the nearby personnel door to the building and looked in but saw nothing. I had sent Joel Lindsey off-range to repair the microwave at Pedro Dome, and I was sure, even though it was uncharacteristic of him, that he was the culprit. I vowed to bring it up at Monday-morning coffee.

Poker Flat technician Carroll Coe preparing for a sounding rocket launch in the blockhouse during the winter 1977 launch season

I walked across to my room at Poker Inn, stripped off my sweaty clothes, and walked down the hall to take a shower. I redressed and drove a couple of miles back toward Fairbanks to the Chatanika Gold Camp, where I treated myself to an expensive dinner around 6 p.m. After dinner, I drove down to Ron and Shirley Franklin's Chatanika Lodge to have a drink before returning to Poker Flat. This was very unusual—I rarely had a drink to be sociable. As I walked into the lodge, Ron immediately asked if I knew what was going on. When I asked why, he said two or three state trooper vehicles had driven by with their sirens screaming and had stopped somewhere near Poker Flat.

I immediately drove back to Poker Flat. I saw an unmarked state trooper vehicle near the front gate. I stopped and talked to him briefly and learned that they had been called out because of a robbery. As I pulled up in front of Poker Inn, a state trooper put someone in a vehicle and drove away.

Another trooper came over with Eldon Thompson and asked if I could tell them anything about the gray truck. The pickup bed was heaped with things: soda pop, fire axes, etc. The rear seat was filled with Poker Flat T-shirts and other odd stuff. I told them that none of the material had been in there when I had seen the vehicle in front of the Payload Assembly Building only a couple of hours earlier.

It seems Eldon had heard some sort of noise and walked from his trailer house toward Poker Inn, where he saw two guys loading stuff into one of our gray trucks. When he asked what they were doing, they mumbled something that led him to believe they were part of the Poker Creek watershed project. But they were behaving very strangely. He walked on by and all the way out to the front gate, where he called on the gate phone back to his girlfriend, Parlee, in his trailer house. He had her get his pistol and drive out to the front gate, warning her to stop for no one. On the way back in, Eldon did not see the two young men, and the pickup had been moved.

He sent Parlee back to his trailer and set out on foot with his pistol. He caught up with the two fellows near the back gate near pad 4. They had gotten stuck on flat, greasy ground and appeared to be trying to put the vehicle into four-wheel drive. When Eldon yelled out to them, one of the fellows ran off into the brush and the other one stayed. He was very young, in his teens, and scared to death as Eldon pointed a pistol at him. Eldon walked him back to Poker Inn. He had Parlee come over to Poker Inn, hold his pistol on the young buck, and call the State Troopers for help. He walked back and got the pickup truck with its load and drove it back to Poker Inn.

Parlee in the meantime had called the State Troopers and told them she was holding a gun on a burglar. The troopers misunderstood her, probably because of our characteristically noisy telephone lines, and thought someone with a gun was holding her hostage. They had just had this situation a week before out in Ester, Alaska, so they donned their flak jackets and sped toward Poker Flat.

When the troopers arrived and understood the situation, they moved about until they discovered a car that looked like it belonged to these two young men parked alongside the road just outside of Poker Flat's front gate. They watched the car and waited for the mosquitoes to drive their quarry out of the brush, where they were sure he was hiding. Sure enough, they got their man a half-hour or so later.

He told them that he had just gotten out of jail in Anchorage. He was about 24 years old. He had picked up his 17-year-old companion in Anchorage. They had stolen the car and were headed to Circle, Alaska, where they were going to homestead. They spotted the "abandoned military site" (Poker Flat) along the road and took only the things they needed to homestead. He and his young friend were stoned.

Eldon and the troopers left, and I stayed at Poker Flat. As I looked about, I could see that they had first pilfered my desk and took a sack of peanuts still in their shells. They left a trail of peanut shells in the office building, the Payload Assembly Building, and my room in Poker Inn. They had stolen a tape recorder, which they accidentally turned on. Their conversations were very stilted: "Wow, look at this, we'll really be able to use that." I think we gained an axe or shovel that we never had before. I lost a small vial of gold dust in water I had gotten from a friend's Yukon Territory mine in 1964.

The teenager had run afoul of the law in Louisiana, where he lived with his parents, and had been shipped to live with a sister in Anchorage. The State of Alaska paid his way home and did not prosecute him. The 24-year-old came to trial later, and I testified to what I had found. This was his third offense, and he drew a long sentence.

The Golden Days Parade

In the spring of 1979, I made plans to build a float for the Fairbanks Golden Days Grand Parade. The theme was something like "visions of the past, present, and future." I had the idea that we would have the model gold dredge that belonged to the University of Alaska Fairbanks on the gooseneck trailer to represent the past. Behind it I planned to build a corrugated-tin-roofed, open-sided building to represent the Old F.E. Gold Camp. Behind this I had the range crew mount the T-9 radar and a launcher with a model Loki-Dart to represent the present. At the rear I mounted two sections of triangular radio tower, and across their top I put a Nike Hercules radar fiberglass cover. R2-D2, Darth Vader, and Princess Leia would stand under this to represent the future.

I enlisted Bob Rolle, who helped his mother, Trish Galleto, run the Old F.E. Gold Camp. I told Bob I would take care of getting the float trailer ready, but we needed people from the Old F.E. Gold Camp dressed in old-time costumes, including some women dressed up as cancan dancers. They would ride and dance under the corrugated roof section. Bob arranged for a local trucker, Al Macquaid, to pull our float with his truck tractor.

We had a really long, and really heavy, US Air Force surplus fifth-wheel trailer at Poker Flat. I had the range crew pull it out of storage and park it between the Payload Assembly Building and the office building. Right from the start I planned the layout of the float to protect the people riding it in case it rained the day of the parade, which it often did. The corrugated roof would protect the Old F.E. Gold Camp crew. Darth Vader, Princess Leia, and R2-D2 would be under the protective radar dish shield at the rear. The rocket man, wearing clean coveralls with a Poker Flat hard hat, could duck under the corrugated roof. I strung some snow fencing around as a skirt on the trailer and hung some pictures and signs there too. I put a model rocket and *Star Wars* X-wing fighter on each side. On the rear I mounted the four-foot-diameter base plate for an abandoned antenna. I painted it gold and made it a large representation of that year's Golden Days logo pin.

The gold dredge model for the float was another project. When I first came to the University of Alaska in 1963, a model gold dredge sat in the middle of the fountain on campus. Employees of gold mining dredge operations had built it. In 1967 I saw it again at Alaskaland in Fairbanks.

In 1979 I called Earl Beistline, a mining engineering professor at the university, and told him that I wanted to use the dredge as a part of a Golden Days parade float. Ever enthusiastic, Earl was tickled that it would appear in public again and explained apologetically where I would find it. Two years before, mining engineering students had recovered it from Alaskaland and brought it to the University of Alaska Physical Plant, where they took it apart to repair it. They got it apart, but never fixed it. I found it lying in overgrown grass along the south side of the Physical Plant. I loaded it up on our old green Dodge 1.5-ton stakebed and brought it to Poker Flat. It was so heavy it made the truck sag.

Left to right: Neal Brown's children Steve Sweet, Nathaniel Brown, Melody Brown Burkins, Michael Sweet, and Kristopher Brown, with R2D2 in 1980

At Poker Flat we put it up on top of the trailer's gooseneck. I had an old antenna rotator motor and adapted it to run the bucket line. In the last days before the parade, I got the idea that we would have two Civil Defense water buckets hooked together by a pipe near their bases sitting underneath the exit of the two sluicing chutes. I bought a sump pump and ran clear plastic three-quarter-inch tubing up into the area the bucket line dumped into. When we turned on the electricity, the bucket line went around as if digging muck and bringing it up into the area where water was mixed with it. We glued some large rocks onto the side sluice chutes and painted them to look like big gold nuggets. Everything about the float looked and sounded great. We added a gasoline-powered generator to power the electric motors.

We covered the generator with a mockup of the iron cookstove at the Gold Camp. I made this from some metal-clad quarter-inch plywood that we painted to look like a cookstove. We even included a chimney for the generator exhaust. However, we did not test this setup thoroughly.

I did much of the work myself evenings and weekends. Poker Flat employee Charley Lasater pitched in at one point, and Jim (Andy) Andrews, the wind-weight guy from Wallops, got intrigued and repainted the gold dredge with silver paint.

As the parade day approached, I needed a Princess Leia. Randi Wagner was not available or interested but loaned me her outfit. At nearly the last minute I talked Penelope (Noecker) Dufseth into wearing the outfit. Penelope was nervous until we got moving, but then she really got into it. I was wearing my Darth Vader getup I had made for Halloween the year before, because my children and I were all super Star Wars fans.

Just before the start of the parade, all of the floats were parked on the exit road leading off Fort Wainwright toward downtown Fairbanks. We had to be in position there by 9:30 a.m. The judges would come by about 10, and the parade itself would start about 11. We cranked up the generator that powered the gold dredge just before the judges came by.

The floats in front of us started moving right on time. As they moved, various people or horse groups moved in between the floats. As we turned the corner from Gaffney to Noble Street, a small group of four or five young people moved in front of us, carrying a banner that stretched across the street. We did not know what that meant.

The electrical generator overheated about this same time and quit. No more moving bucket line or water flowing from the model gold dredge. People aboard our float were asking the crowd along the street what the banner said. Most people thought we already knew and just clapped and shouted their approval. Finally, someone got off our slow-moving float and ran forward to see what the banner said: "Winner of the Governor's Award, Top Prize for the 1979 Golden Days Parade." We were ecstatic.

We traveled up several crowd-lined blocks until we got to the judges' stand, which was on the east side of Noble Street opposite the Northward Building. They stopped our float and offered us the two-foot-tall champions cup. Everyone insisted that I accept it, and I carried it for the rest of the parade. The parade route turned from Gaffney onto Second Avenue and along the main two blocks of downtown Fairbanks. We then turned south onto Cushman, which was lined with people for another three or four blocks. All along the parade route, crowds roared their approval for our float. The official parade route ended as we turned off Cushman onto Airport Road.

Well-wishers gathered around. We celebrated our success with lots of hugs and with friends who had come to congratulate us. Both UAF and Geophysical Institute administrators thanked us for our efforts. This began a tradition of Poker Flat having an entry in the Golden Days Parade each year.

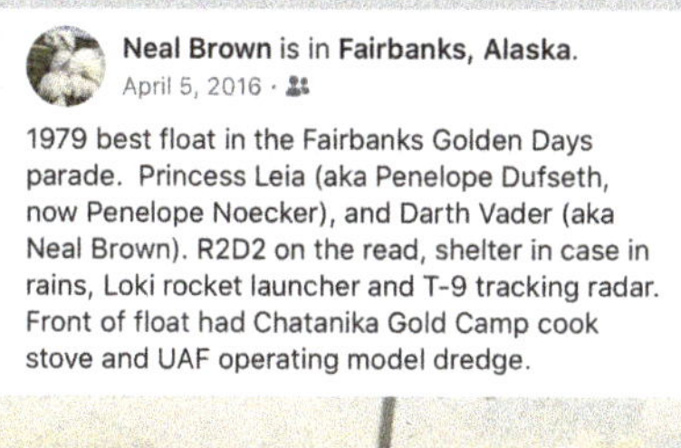

Neal Brown is in **Fairbanks, Alaska**.
April 5, 2016 ·

1979 best float in the Fairbanks Golden Days parade. Princess Leia (aka Penelope Dufseth, now Penelope Noecker), and Darth Vader (aka Neal Brown). R2D2 on the read, shelter in case in rains, Loki rocket launcher and T-9 tracking radar. Front of float had Chatanika Gold Camp cook stove and UAF operating model dredge.

Chapter 8
The 1980s

C-Band Radar Foundation, 1980

NASA Wallops made plans to replace their S-band VERLORT radar system with their version of the national range C-band radar system. They developed and sent us plans to fabricate a permanent foundation for it. They would eventually build five of these C-band radar systems, with three being temporarily deployed around the world and two always at Wallops being overhauled and upgraded with the latest internal instrumentation.

We looked at NASA's requirements for location, stability, and position above the local landscape and waited for their in-house design engineers to provide us with engineering drawings of the structure itself. We were to review the drawings and submit a budget to build the structure on our soil foundation. When the engineered drawings arrived, we investigated what it would cost to buy the materials they specified and fabricate the structure they proposed. From past experience, we knew that NASA Wallops engineers had a tendency to fine-tune their designs to meet the requirements in the most economical fashion possible.

We, on the other hand, had no one on our staff who fully understood the strengths of our soils to provide a foundation for the proposed radar pedestal, or who could evaluate their engineering design for our situation. What we had were a bunch of experienced hooligans with Alaska homesteader personalities who had a hunch they could do it better and cheaper than "those NASA guys."

N. Brown, *Northern Lights and Rocket Flights*, Springer Biographies,
https://doi.org/10.1007/978-3-032-14598-7_8

We responded with a crude design of our own that used 48-inch-diameter pipe with half-inch wall thickness left over from the construction of the Trans-Alaska Pipeline. Without doing the numbers, I knew it exceeded the structural specifications for the 18-inch thin-walled pipe NASA had requested, and there was so much of it in the Fairbanks area we could buy it as scrap iron. Without engineering, we knew we needed to dig 8 or 10 feet of muck out from where the pedestal would be and replace it with good-sized gravel from nearby piles of gold mining dredge tailings. Best of all, doing it our way would cost about $5000 for materials versus $18,000 the NASA way. They gave us the go-ahead. I learned much later they remained a bit worried that our design would not be stiff enough.

We had time to do this job right, and we had it finished and painted by the time a trucking company delivered the two trailers that made up the NASA C-band radar system. A NASA radar contract crew showed up soon after and told us the radar pedestal looked better and easier to move around on than any they had ever seen.

It took several days to mount the radar dish pedestal onto the foundation we had built and to check out and power up the C-band radar in its self-contained trailer. The radar contractors checked the stability of our foundation pedestal several different ways, using the sun as a signal source and using specially equipped Earth-orbiting satellites. They communicated with NASA radar engineers at Wallops about their progress. Within a few days, we received praise from NASA radar and structural engineers for the exceptional stability of the pedestal we had built.

Red Challenged

I knew what I wanted: a photograph of the display Nike-Tomahawk rocket with payload that stands at the front gate of Poker Flat Research Range with a vertical red cloud of rocket-borne lithium vapor at the same time. In 1982 I overcame two major obstacles to get it.

The first obstacle was that I cannot see the 670-nm red light of lithium vapor because of my red-green color blindness. But I knew from viewing previous rocket-borne releases that I could see the tiny glow of it being dispensed through the side of the rocket as the rocket fell back to Earth. The second obstacle was that I had to lie in a puddle of icy water to take the picture.

Time-lapse photographs of the vertical cloud of lithium vapor dispensed by a falling sounding rocket are used to determine wind speeds and directions. My goal was just to obtain a cool picture for public relations use, not to collect scientific data. I found an opportunity with the launch of a sounding rocket on April 16, 1982. This rocket carried and dispensed barium and lithium vapors as it flew. I stood at the entrance gate as it lifted off early one morning and wondered just how cold I was going to get lying in that 2-inch-deep puddle of water that was right where I needed to be to take the picture. I had no choice because the only camera tripod I had was about 6 inches tall, and I had to aim the camera so the rocket and the trail release were together in the camera frame. Then I had to manually hold the shutter open during the seconds the lithium vapor would be coming out of the rocket as it fell.

The initial shock of cold was the worst. After that my body heated the water next to my skin, much as happens in a diver's wetsuit. I settled into a cramped position looking through my camera viewfinder up past the gate rocket to the sky overhead. Then I saw a small bright light moving slowly downward and knew it was part of the lithium coming out of the rocket. I got both into the camera frame and opened and held open the shutter until the bright light disappeared.

I was confident that I had gotten the picture I wanted, even though I had not seen it in real time. I packed up my camera and went back to Poker Inn to change clothes. Two weeks later, I was rewarded with seeing for the first time the picture I had successfully taken. As I knew it would, the photographic process had turned the red light of lithium into a red that I could see. It remains one of my favorite unique pictures from my 18 years as director of Poker Flat.

Front Gate Display Rocket

Gil Moore of Thiokol called me in the mid-1970s and asked if I wanted the Nike-Tomahawk rocket model that had sat outside their office in Ogden, Utah, since they were moving their offices. I said yes, and Gil had Thiokol employees pack it neatly into boxes and paid the cost to ship it to Poker Flat.

We already had a complete Tomahawk fin shroud assembly with fins attached. NASA had launched a recovery payload Nike-Tomahawk mission a few years earlier that had landed in the Arctic National Wildlife Refuge in the Brooks Range north of Fort Yukon, Alaska. Because it was the Refuge, we were obligated to remove everything from this mission we could find. To our great surprise, we found the payload on a windswept lake and the Tomahawk nearby. The Tomahawk head cap had augured into the ground on impact, leaving the rest of the Tomahawk standing. The fin shroud had broken loose and slid forward with the fins intact.

When the parts arrived from Utah, we welded a piece of steel pipe into the Nike nozzle and mounted the Nike fins and the interstage adapter that mated the Nike to the Tomahawk. We put a fresh coat of paint on our new display rocket, dug a hole in the ground by the front gate, and planted the complete Nike-Tomahawk there, where it still stands.

NORAD Space Command Calls

NORAD Space Command found us on April 1, 1983. We were counting down rockets the night of March 31, 1983, without success. I drove the 35 miles home and had just gotten to sleep when my assistant and launch director Pat Young called from Poker Flat, and said I needed to come back right away and talk to someone from NORAD. NORAD's attention had been caught by the March 19 launch of a Talos-Castor for Earth Limb Infrared Auroral Studies (ELIAS). The caller had been tasked with learning about the sudden appearance of a strange rocket burn and had heard from the North Pole Police Department that we had been launching rockets. He initially talked with Dave Behr, who was on heater watch, keeping rockets on launchpads warm. An air force captain identified himself and demanded to talk with whoever was in charge of the facility right away. Behr told the captain that the whole crew had gone home after spending all night trying to launch a rocket. The captain kept insisting that he speak with someone, so Dave reluctantly called Pat, who was the launch controller and lived in a house on range as our year-round site caretaker. I drove groggily back to Poker Flat, and Pat and I called NORAD.

Neal Brown's photo taken in 1982 of the Nike-Tomahawk sounding rocket at the front gate of Poker Flat with a red cloud of rocket-borne lithium vapor behind it

The captain informed us that NORAD had no idea who we were and that certain things about that launch had greatly concerned them. When I told him that we coordinated all launches a year in advance with the Alaska office of the FAA and that they communicated not only with the military in Alaska but also with the NORAD office at Elmendorf Air Force Base, he refused to believe me.

I got an inkling that an orbiting Department of Defense satellite had seen something. I politely told him that we had been launching rockets since 1969 for a number of US government agencies. I also told him that an air force KC-135 scientifically instrumented aircraft had been orbiting the launch of a Talos rocket motor-boosted scientific experiment precisely *because* the Talos burn resembled the infrared signature for an ICBM launch. The problem was that the fast-burn boosters of the ICBM did not give US observing assets enough information to determine where the payload might be going.

The captain insisted that we must call him at NORAD before any launch and keep an open hotline with NORAD during launch attempts. I politely told him we had no money to do anything like that, but if he sent us a requirements document and money to set up the line, we would be glad to do it. A few days and telephone calls later, NASA told us they would pay us to establish a hotline with NORAD.

A few weeks later, I received a letter inviting me to attend a Launch Control Users meeting at NORAD headquarters at Peterson Air Force Base in Colorado Springs, Colorado. Pat and I joined about 40 other range representatives for this meeting. We quickly learned that the criteria of notifications for launches varied drastically from range to range. For example, safety at Cape Canaveral in Florida was all about protecting astronauts. Safety at Kwajalein in the South Pacific was all about launching when no Soviet spy satellites were close enough to observe what was happening.

NORAD wanted us to avoid accidentally hitting a Russian spy satellite, which would cause an international incident. Our rockets sometimes reached apogees of over 500 miles, which was well above the orbital altitude of several satellites. I told them we had helped NASA do some calculations to make sure that the chemical releases from satellite programs did not coat another NASA satellite unnecessarily with barium or lithium. They attempted to put us on the hot seat about protocols. Ostensibly, by international agreement, spacecraft orbits are announced and made available worldwide to everyone within 2 days of launch. When I said we did nothing, air force lawyers present were astounded. When they tried to pin me down about how we knew we were not going to hit a satellite, I countered by asking if NORAD updated the orbital parameters for the US spy satellites, which often changed orbit to look at something in particular on the ground. I watched a couple of US Air Force lawyers go into a huddle and then come back to our group and not say anything more about the subject.

After 2 days of learning about who did what and how they did it, we packed into buses for a tour of the NORAD threat assessment facilities in Cheyenne Mountain, just west of Colorado Springs. We got out of the buses and walked through that amazingly huge polished bank vault door you see in the Hollywood movies. The NORAD systems inside Cheyenne Mountain were designed to survive the direct hit of an ICBM equipped with a nuclear weapon. Some distance inside, we were ushered into one of two large manned instrumented control centers mounted on gigantic coil springs. We walked across an elevated ramp into one of two threat assessment facilities. We had a grand tour of a relatively small, highly organized, and functional control center that would be used to acquire, identify, and analyze anything orbiting the Earth.

Quite by accident I asked about a canvas-covered opening in a side wall and was told it was a quick way to get from the Space Threat Assessment facility to the Airborne Threat Assessment facility. A light went on in my head: the FAA in Anchorage was set up to communicate with the NORAD Airborne Threat Assessment component, which looked for bombers flying low into Alaska as seen by coastal radars. The Clear, Alaska, Ballistic Missile Early Warning System radar was connected to NORAD's Space Threat Assessment module, and NORAD's Airborne Threat Assessment systems in Alaska—such as the DEW radars that looked for Russian ICBM launches—were connected to Elmendorf Air Force Base, which communicated to the module next door.

I asked if we could go see the folks in the module next door and was told, "Sure, but watch your step—it's a serious fall if you lose your balance." The safe way to get from one module to the other was by walking out the way we had come and over to the other module. When I pulled the canvas aside, I saw an 8-foot wooden-plank walkway with no handrails that was at least a 8 feet above the area below. The walkway bounced as I walked across.

Once inside the Airborne Threat Assessment unit, followed by a puzzled NORAD man, I asked if anyone in this module had ever heard of Poker Flat Research Range and got an almost unanimous positive response. I introduced myself and Pat, and we had a fine time reminiscing. All they cared about with regard to us was when our rockets flew from the ground to 50,000-feet altitude or when they came back down from 50,000 feet to the ground. They were concerned about the same things the FAA worried about.

This was the first time that the right people in NORAD understood why I had claimed we regularly talked with NORAD via the FAA.

Commemorative photo acknowledging the 200th launch of a sounding rocket at Poker Flat in 1984. Pictured from left to right: Neal Brown, Dan Kush of SDC, Lt. Ed Blasi and Capt. Jim Shuster of of USAF/BMO, Art Buckley of MIT/LL, and Val King, USU

200th Launch Trophy

Launch number 100 occurred on June 13, 1975, and went without recognition. It was a spectacular failure that I wrote about in Chap. 7. Launch number 200 was an outstanding success that we launched on February 2, 1984, for the Ballistic Missile Defense Agency (BMDA). It was a handsome three-stage rocket, a Talos-Sergeant-Hydac.

I wanted to commemorate this launch, so I worked with Larry Kozycki, head of the GI machine shop, to create a chrome-plated model of the rockets and payload and mount them on wood bases with commemorative plaques. I gave them to the people who played a major role in this event:

- Jim Gray, a longtime friend of sounding rockets and employee at NASA Wallops Flight Facility at Wallops Island, Virginia
- U.S. Senator Ted Stevens
- Neil Davis, who created Poker Flat Research Range
- Juan Roederer, then-director of the Geophysical Institute
- Jim Schuster, an air force captain who was program manager for the BMDA effort
- Frank Shannon, NASA radar contractor
- Bobby Wessels, a contractor for NASA telemetry

Command Destruct NASA Aerostat Balloon

One of the more interesting nonrocket science projects we supported at Poker Flat was a set of summertime atmospheric electricity measurements made with the use of a tethered aerostat in 1984. NASA Wallops provided the principal investigator, Bob Holzworth of the University of Washington, with the aerostat balloon and launch and recovery services. The aerostat balloon had fins like a dirigible and was perhaps 80-feet-long and 25 feet in diameter. It was attached by Kevlar cable to a substantial motorized winch mounted on a trailer. A separate insulated electric conductor ran from the aerostat balloon to an electrically insulated cylinder about 3 feet in diameter and 6-feet-long, also mounted on a trailer.

Atmospheric electricity and electric field measurements were made from an instrumentation package that flew beneath the aerostat balloon. Measurements were to be made from near the surface at Poker Flat to as much as a mile high. Since this got into FAA airspace, we had to coordinate our tethered flights with them. NASA Wallops had flown similar aerostat balloons in FAA airspace before and had the mandatory radio-commanded destruct system onboard in case something went wrong, like the tether cable breaking and the aerostat balloon starting to fly away uncontrolled. This time we were ultimately responsible for all flight safety if something went wrong, but of course we all knew if the mission were being done for a federal agency such as NASA or the air force, they would be the deep pocket for financial damages.

I was interested in the details of the radio-controlled command destruct and learned about the radio transmitter and receiver from Wallops staff on site at Poker

Flat. The receiver was on top of the aerostat. The command destruct signal would fire a pin puller that held closed an assembly of flaps on top of the aerostat balloon. Once the flaps were opened, the helium gas would escape and the aerostat balloon would rapidly fall back to the ground.

We flew several missions to make measurements. The electric potential at altitude was communicated by the insulated conductor back to the large insulated cylinder on the trailer. The measurements of electric potential made from this cylinder with respect to ground were the prime data.

Then one day, a thunderstorm developed and quickly moved our way while the aerostat balloon was still a few thousand feet up. We started pulling the aerostat balloon down as fast as the winch would turn without breaking the Kevlar line. As it passed through a thousand-foot altitude, large electric sparks started jumping 3 feet through the air from the insulated cylinder to electrical ground. As the aerostat balloon came down, they kept coming quicker and quicker. This was an indication that we were losing the race to get it down ahead of the thunderstorm. They also created a horrific noise. These minor lightning bolts kept getting louder and brighter as the aerostat balloon approached the ground.

Rain and hail started falling. When the aerostat balloon was perhaps 150 feet above the ground, the insulated electrical conductor burned in two right before our eyes. The Kevlar winch cable appeared to still be holding, but the aerostat balloon was starting to move up again. The insulated electrical conductor had been playing a part in helping pull it down.

I was videotaping this amazing situation and saw the NASA command destruct folks send the radio signal to deflate the aerostat. I could not tell if it had worked, but the aerostat balloon quit rising and started slowly moving down again.

After the rain and hail ended, it was obvious the aerostat balloon was still fully inflated. I talked with the NASA Wallops team about whether it had worked. While the guy in charge assured me it had, another grizzled NASA hand said of course it did not, because it had been glued shut. I had a hunch the grizzled fellow was telling the truth and the other guy was not.

I hung around while they deflated the aerostat balloon, as we did after each flight. We could see that the pin puller that released the flaps had fired, but the flaps remained in position. It turned out that the flap assembly leaked precious helium. The field fix, a couple years before, had been to glue the flaps together. I was told they had never tested the command destruct system with this fix in place to see if the flaps would release when the pin was pulled. This was only one of several situations that led me to question whether NASA really followed its own safety rules.

The Aurora Color Television Project

Early monochrome cameras, although very sensitive, missed much of the beauty and scientific content of the colorful aurora. Neil Davis brought the first low-light-level black-and-white television system to the Geophysical Institute in 1964. He brought then-engineer Tom Hallinan with him to operate the system. In 1972, Tom assembled a color television system by combining three of the monochrome cameras that were appropriately filtered for the three primary colors. This system may well have been the most sensitive color television camera ever built up to that time, and it allowed the color recording of even the weakest auroras. However, because it depended on three separate cameras, each with its own sensitivity control, it was not possible to calibrate the system or to maintain an accurate color balance.

Further development of the system came in the late 1970s, when professional television production teams from NHK of Japan contacted Syun-Ichi Akasofu, then director of the Geophysical Institute, for help in acquiring high-quality, color moving images of the aurora. This led to cooperative work in developing a broadcast-quality, low-light-level color television camera. Tom worked with several Japanese groups to help them develop a camera sensitive enough to photograph the aurora. PAVIC, a production company, wanted high-quality television images of the aurora for a special production about Frank Yasuda, a Japanese-born Alaska pioneer. In 1908, Yasuda saved his shipmates' lives by walking to Point Barrow for help after their vessel was caught in the pack ice. He subsequently stayed in Alaska and became one of its noteworthy and leading citizens. Much of the testing of these experimental cameras was done at Poker Flat with the support of Dan Osborne, the optics site engineer.

In 1983, Tom wrote a successful grant request to the National Science Foundation and NASA for $120,000 to fund the purchase of a PAVIC-type camera for use in auroral research. In mid-March of 1984, we took delivery of a color television camera capable of registering the colors of aurora on videotape that could be played on any television. That spring, Dan managed to catch a beautiful display of sunlit aurora and six rocket-borne releases of barium. The data from those releases helped answer questions about the rate of ionization of barium.

In addition to the scientific data collection, we decided to build a collection of aesthetically pleasing auroral sequences. This was a very deliberate effort to obtain images for viewing by the general public. It also involved considerable extra work on nights and weekends during times when the camera was not required for scientific studies. The result was a unique and stunning collection of color auroral recordings. Tom Hallinan and Garry Meltvedt edited some of that video into what we called the "Pope tape," which was presented as welcome gifts to President Ronald Reagan and Pope John Paul II when they met in Fairbanks in June 1984. It was about 10 min long and gave a very brief explanation of the aurora.

By January 1985, I had permanently installed the color camera under its own 5-foot-diameter Plexiglas dome at the optics building at Poker Flat. In addition to our normal work, Tom, Dan, and I operated the camera every moonless night throughout the winter and spring of 1985 to obtain more than 30 h of high-quality auroral video. This spectacular footage was brought to the attention of CBS by local affiliate KTVF, and our video was seen by millions on the CBS *Morning News*.

Neal Brown in 1985 being interviewed by CBS local affiliate KTVF about the aurora borealis

By mid-March we had quite a bit of spectacular video. We had rockets on the rail, and a PBS Planet Earth film crew underfoot at times. By the end of the first week of April, we had launched all of our rockets, taking all the video we could, and the film crew headed back south. I sat in front of a television monitor and made detailed notes on all 18 h of aurora videotape.

By late April, we had several things going, the most notable of which was a report to NASA and NSF. We were all too busy to write up a final report, so Dan said he would make a video and send it with a cover letter. I selected the best scenes, which Dan further narrowed down to show scenes from a quiet evening through an auroral breakup to pulsating morning to make a half-hour video. He used a local video production company's editing suite to put it all together. The result was 24 min of the best aurora footage we had. I insisted that it be accompanied by music, gave him my ideas, and stepped back to let him take charge. Dan assembled the scenes into the story line I had requested and set it to the music of *Ursa Major*, a classical music production written by University of Alaska professor/conductor Gordon Wright and performed by the Fairbanks Symphony Orchestra in 1978. We anticipated that this version would be duplicated commercially and sold to aurora enthusiasts, tourists, and fellow scientists. We titled it *Aurora*.

Shortly thereafter, the Scientific Advisory Board of the Geophysical Institute convened in Fairbanks for its annual review. During a break in their busy schedule, they viewed a 10-min section of *Aurora*. That 10-min filler grew to almost an hour when the board insisted on seeing the entire production and proceeded to formulate a plan for us to bring the wonders of the aurora to the public. The advisory board specifically emphasized that we three should possess the copyright in our names and thereby retain control of the uses of the program and raw footage. Otherwise, the images would enter the public domain and be available for reproduction through a variety of inadequate reproduction techniques.

On the basis of these discussions with the advisory board, we formed the Aurora Color Television Project (ACTP) as an organization within the Geophysical Institute. So that it would be self-supporting, ACTP sought to maximize commercial returns from sales of *Aurora* and from licensing use of the footage, with all proceeds going to the Geophysical Institute. In return, ACTP would operate under the umbrella of the Geophysical Institute, with access to its facilities. ACTP would retain control of the proceeds, subject to the requirement that the money would be used to further science and education and be spent in accord with State of Alaska and University of Alaska procedures. All expenses would be approved by the director and the business manager of the Geophysical Institute. We never drew a single penny of salary from the proceeds.

The success of this approach was demonstrated by more than 62,000 copies in public hands, by the many instances in which the aurora has been used in major television productions and features with full credit to the University of Alaska Fairbanks, and by the more than $200,000 in revenue that the project generated for the Geophysical Institute. Perhaps the largest single audience viewed the aurora when Charles Kuralt used some of the footage for a short feature on the aurora during the 1994 Lillehammer Olympics. Each time the videos are played, logos and credits promote the science education outreach efforts of the Geophysical Institute of the University of Alaska Fairbanks, NASA, and the National Science Foundation.

Funds from the sales of *Aurora* have been used to purchase equipment; to replace, supplement, and add capabilities to the research pursued by the Geophysical Institute; to travel to meetings to learn about recent advances in imaging technology that can be used both in our research and in our efforts to promote public understanding of science; and to fully fund the production of the educational videotape *The Aurora Explained*, which was copyrighted in 1992.

The GI continues to receive requests for video images of the aurora. Licensed sales of auroral images have allowed other national and international producers of educational, news, and commercial television to have broadcasts, sell videotapes, and have public showings of the aurora. Each production has been required to credit the University of Alaska Fairbanks and to properly present the aurora. As a result, many millions of television viewers around the world have had a chance to see the aurora in its true color and motion. Travelers have been motivated to come to Alaska to see the aurora, the public has been informed and entertained, and students have been inspired to learn about the aurora.

Radio Moscow

In 1985, Utah State University built a rocket payload to make high-speed measurements of electron density for the Wideband II program. They built a small radio-frequency oscillator and coupled it into adjoining sections of the payload. The electron density data was derived from an antenna impedance measuring system that responded to the electron density adjacent to the rocket payload as it flew. The system was so sensitive that the steel of the launch rail kept us from knowing whether this part of the payload was working properly once the payload was installed on the launcher. But the device was simple in concept, and the design had flown many times before, so everyone assumed it was working.

A couple seconds after launch, the data from this part of the payload became a mystery that puzzled everyone involved. The instrument seemed to be working, but the data wavered in ways uncharacteristic of electron density as a function of altitude above the surface of the Earth. For the next few days, several people pondered the data and attempted without success to explain what had happened.

About 3 weeks later, long after everyone involved with the launch had gone home, Dean Feken, who worked full-time for Space Data Corporation taking care of the telemetry site at Poker Flat, played the data back through a radio receiver and found he was listening to Radio Moscow, a powerful Russian radio broadcast.

The scientists and engineers at Utah State University had had problems finding a radio frequency free of radio signals and had decided to use 10.7 MHz. This frequency had been set aside by many nations of the world for use as an intermediate frequency used internally in various radio receiver signal-processing systems. Much of the signal-processing detail takes place at this frequency in radio receivers, no matter what frequency they are designed for. In Logan, Utah, and during ground tests at Poker Flat in Alaska, 10.7 MHz was fine. But Russia had not honored that intermediate frequency set aside and had at least one broadcast transmitter on that frequency. When the rocket payload cleared the local terrain, the sensitive radio system onboard heard Radio Moscow, to the exclusion of everything else.

Science Education: Key to the Future

In 1987, Gunter Weller asked me to head up the Arctic Science Conference for the American Association of Science, of which I had been a member for years. I told him I would do it if we could make it about science education, involve Fairbanks teachers, and give them a discount on the entry fee. He secured permission for me to go ahead. I had roughly $50,000 available for public relations, printing, and honorariums for invited speakers. None of the other professionals involved got paid except Gunter's administrative assistant, Cindy Wilson.

I contacted several Fairbanks teachers, who hoped I could get the Fairbanks North Star Borough School District to declare it a teacher in-service event so they would be able to attend and get paid. I contacted Helen Barrett, who headed that up, and she agreed.

The Fairbanks North Star Borough let us use Ryan Junior High School and the large theater at Lathrop High School next door. I had chosen the theme for the conference, "Science Education: Key to the Future," and drew what would be our logo of Alaska and western Canada like a gigantic padlock with a large key with "Science Education" embossed on it. Roughly 400 professional scientists came from Alaska and western Canada, and 1,200 Fairbanks schoolteachers joined the 3-day event.

We wanted to have two science teacher educators give us two different views on how teachers should be taught. My friend Doug Schamel had worked with Verne Rockcastle from Cornell University and suggested we ask him to be our moderator for a discussion between Arnold Irons from the University of Washington and the head of the National Science Teaching Association (NSTA).

Arnold Irons was a theoretical physicist who developed a major science teacher program at the University of Washington. He was a by-the-book teacher, insisting that teachers learn some basic physics equations such as force equals mass times acceleration. The NSTA representative was much looser, saying that was too complex for teachers and that they should use language and words that fit better with the world of hands-on references. They came together in the Lathrop auditorium in the last few hours of our conference. Arnold and the NSTA man were nice to one another but presented their different points of view eloquently, and Verne made sure they got equal time. Fairbanks teacher Judy Schiffler told me Arnold made them realize they should have learned some basic science earlier in their lives.

I sat in on one of Verne's presentations with kindergarten teachers. He was lying on the floor with them with rubber bands, crayons, and small slabs of wood showing them hands-on ways to demonstrate some of the laws of physics to little kids.

Outhouse Races

The first annual Chatanika Lodge outhouse race was held in late March of 1980. Our range mechanic, two other range people, and I built a lightweight good-sized replica of an outhouse on skis, and invited Sue Royston from the Geophysical Institute Business Office to be our official rider.

We had acquired several military surplus parachute-drying towers and used their parts to construct our frame. We covered it with lightweight highway plastic fencing. A toilet seat was mandatory, so we mounted it securely in the frame with a five-gallon plastic bucket under the hole. Our outhouse was also one of the few that had handholds for our rider. We had Sue test the outhouse seat, took her to see the terrain we would run, and offered to install a seat belt, but she thought she could hold on just fine. A wood or metal roof was required, and ours was lightweight.

A makeshift trail was blazed through the trees down a steep, snow-covered hill from the starting point at the Chatanika Gold Camp to the Steese Highway. Then a packed trail several outhouses wide ran alongside the Steese Highway to the finish line at the Chatanika Lodge.

Twenty-five outhouse teams competed in this first-ever race. Most of the contestants were regular clientele of the Chatanika Lodge, a popular weekend out-of-town bar that featured food and music. It sat alongside the Steese Highway between the Chatanika incoherent scatter radar facility and Poker Flat.

Each team left the starting line at the Chatanika Gold Camp, a high-end restaurant and boarding house, on a signal whose time was radioed to the Chatanika Lodge. Each team was racing for the fastest time to cover the approximately half-mile distance.

We were a team of old guys in comparison to the other contestants, and we hoped not being drunk like our competitors would work in our favor. We also wore rubber ice grips on our shoes and did not see anyone else who was as smart as we thought we were. We could see we were going to win this race with our structurally sound lightweight frame. We had constructed a well-thought-out outhouse and had the most lightweight rider of any outhouse that day.

We left the starting line in good form but barely survived the beatings we took as we went down the slippery, uneven descent to the Steese Highway. Once alongside the Steese, we muttered a chorus of "Are you OK?" to each other and energized ourselves into synchronizing our push.

We were a mess. Our mechanic was winded after a hundred yards, our ice cleats were flapping loose and getting in the way of a good grip, and the outhouse was being violently jerked around. But we kept going and were happy just to make it to the finish line. As we crossed, we received a huge round of applause from the crowd, which toasted our and every arrival with a bottle of beer or other drink.

We almost won and were barely beaten in total elapsed time. Our outhouse arrived at the finish line in as good a shape as it had begun the race. Most of the other outhouses had disintegrated and lay in the trail for others to avoid. A few came across the finish line with pieces or their mandatory rider missing and were disqualified.

We politely joined others standing about watching our competitors arrive, but none of us were interested in drinking a beer or anything else. We spent most of our time telling one another how this or that muscle hurt and how happy we were that we actually made it to the finish line.

Two years later, the Poker Flat crew won the Chatanika outhouse race with that same outhouse. All of the Poker Flat pushers were younger than we had been, and they actually trained pushing the outhouse on snow-covered range roads. They also knew better than to wear ice cleats on their shoes.

Poker Flat's contribution to the first annual Chatanika Outhouse Race in 1980

AT THE MUSEUM

Neal Brown: A legacy of science exploration

By Aelin Allegood

Neal Brown, a distinguished scientist and educator, dedicated more than four decades to the University of Alaska Fairbanks and inspired countless individuals with his passion for discovery and education. As a UA Museum of the North Education Affiliate, Neal led family rocket programs and guided development of aurora programs for many years. His impact at the museum continues today through the Neal Brown Museum Education Support Endowment.

Born in 1938, Brown's journey led him from early work with NASA to becoming a beloved figure in Alaskan science education. After earning his bachelor's degree from Washington State University in 1961, Brown commenced his career at NASA, focusing on the complexities of atmospheric re-entry for spacecraft. Neal's fascination with the aurora borealis drew him to Fairbanks in the mid-1960s, where he pursued a master's degree in geophysics at UAF. This marked the beginning of a lifelong commitment to understanding the aurora, earth's atmosphere, and space.

PHOTO COURTESY OF FRAN TANNIAN

In July 1980, Neal Brown, Fran Tannian, Michael Sweet, Nat Brown, Steve Sweet, Melody Brown Burkins and Kris Brown posed with their dog Puddles in front of the entrance to Poker Flat Research Range where Neal was the director for 18 years.

At UAF's Geophysical Institute, Brown's career and outreach flourished. He served as the director of the Poker Flat Research Range for 18 years, overseeing numerous research projects and rocket launches. Later, he led the Alaska Space Grant Program, promoting education and research in aerospace and earth sciences across the state. In recognition of his significant contributions to the university and Alaska, UAF awarded him an honorary doctorate in 1997.

During his time at UAF, Neal was an integral part of museum outreach. He served as a content expert and also directly led activities at the museum like rocket building programs for families. UAMN Director of Exhibits, Design and Digital Media Roger Topp says, "I was new to the museum when I was first introduced to Neal, and he was massively influential in how I now see myself within informal public education. He believed in and caused me to believe in the phenomenon of the museum experience, and giving it every ounce of creativity we have, this infinitely inventive space for students and teachers." Neal was celebrated for his ability to ignite curiosity in learners of all ages. His enthusiasm for the aurora was infectious. He frequently delivered lectures and workshops, demystifying the northern lights for audiences far and wide. In his own words captured on video, "I give so many

LEGACY » D4

UAMN PHOTO

Neal Brown, pictured here teaching a model rocket building program at the museum in 2003, supported museum programs that encouraged people of all ages to learn and discover.

LEGACY

Continued from D2

talks to people about the aurora. I tell them how to photograph it. I tell them about what is going on. Many times I tell them, 'Don't even worry about it, just go out and look at it.'"

His family seeks to honor his legacy by fostering the same sense of wonder he instilled in others. His son, Kris Brown explains, "Following the death of my beloved father Neal Brown in 2021, I and my brothers Steve, Nat and Michael, as well as my sister Melody and stepmother Fran, decided to honor him and his work with the Neal Brown Museum Education Support Endowment, a unique collaboration between the Museum of the North and the Geophysical Institute to promote his passion for learning and instill in others his deep curiosity and interest in Alaskan science." Kris reflects on his father's joy in education: "My father loved nothing more than to watch the eyes of a young child, senior, or anyone just light up as he explained the wonders of the aurora." The endowment honors the memory of Neal by providing

The Sunday March 2, 2025, *Fairbanks Daily News-Miner* featured an article about Neal's life and work. After his time at Poker Flat, he led the Alaska Space Grant Program and was involved in outreach and education efforts at the University of Alaska Museum of the North support for education and public programs at the museum. The focus is on science outreach in the community, including partnership with the Geophysical Institute.

Fran Tannian, Neal's wife of 40 years, shares, "As a father and husband, Neal brought great curiosity into my and our children's worlds. Neal gave me a spark to my life that I would not have had without him, and I know also enlightened the lives of all of those around him. And I love this Education Support fund because his memory will continue to give joy and promote curiosity for generations to come."

The endowment's inaugural initiative, "Spring Break Science," is set to launch this March at the UA Museum of the North. Designed to engage families in a flexible way, the program offers a variety of science-themed activities throughout the week. Participants can access a free activity guide, available for download or pickup, featuring at-home and outdoor exploration ideas. The museum will host daily drop-in sessions, showcasing science films and providing materials for self-guided activities such as scavenger hunts, scientific sketching, snow goggle design, and aurora art projects.

The "Spring Break Science" event perpetuates Neal's commitment to making science accessible and exciting for all. Through the Neal Brown Museum Education Support Endowment, Neal's passion for learning and his deep curiosity about Alaska's science will be shared with generations to come.

Explore more

Take part in Spring Break Science from March 10-14. A detailed schedule and the activity guide will be available on the museum's website, bit.ly/springbreak-science.

Family programs from the UA Museum of the North are focusing on moose during March. Early Explorers, for children 5 and younger with their caregivers, will be held on March 7. Siblings are welcome.

Family Day: Moose will be held at the museum on March 22 offering a chance for all ages to explore. Kids 17 and under are admitted free on Family Days. Learn more at bit.ly/uamnfamilydays.

Teens are invited to register for the ARTSci Teen Workshop: Ice scheduled for March 20. Explore the connections between art and science. Info and registration at bit.ly/uamnteenworkshop.

The museum's winter hours are from 9 a.m. to 5:30 p.m., seven days a week. For more information about the museum's collections, programs and events, visit www.uaf.edu/museum or call 907-474-7505.

•••

Aelin Allegood is the development officer for the University of Alaska Museum of the North.

Neal with some kids at Space Camp. Credit University of Alaska Museum of the North

Chapter 9
Neal Leaves Poker Flat

Driving the Steese Highway to Poker Flat

I used to jokingly tell people that over my 18 years of driving the nearly 60-mile round trip each day to Poker Flat, I likely had come close to driving the distance to the moon and back. I had always loved that daily commute. Where else in the world could you work at a space-age facility and enjoy such beautiful scenery and occasional bear, moose, or fox as you drove back and forth to work each day?

Neal Brown Road

At milepost 30 on the Steese Highway north of Fairbanks, you will see an entrance road leading off to your right with my name, Neal Brown, on the signpost. This was a parting gift to me by the staff at Poker Flat when I retired in June of 1989. Earlier in 1989, the Fairbanks North Star Borough had announced they were setting up a 911 emergency response system. In this effort they wanted to eliminate duplicate road names and be sure that all roads had unique names so that emergency responders would know where to go. Dan Osborne and Mary Farrell contacted the borough and received formal approval to name the entrance to Poker Flat Neal Brown Road.

I first learned of the designation when I saw the new signpost with my name on it as I drove to my retirement party. I felt humbled that the people who had worked with me for so long, and whom I cared about so much, felt this way about me.

N. Brown, *Northern Lights and Rocket Flights*, Springer Biographies,
https://doi.org/10.1007/978-3-032-14598-7_9

Sabbatical 1989

When I took the job at Poker Flat in 1971, I said I would stay only 4 years. At that time, I was an up-and-coming young scientist, and I thought if I stayed for more than 4 years, it would be too hard to get back to doing the science I loved, the study of the aurora borealis. I had a great time and loved what I did for 18 years. But eventually I decided I wanted to play—I mean work—in other ways.

I asked Charlie Barth of the University of Colorado in Boulder if he would invite me to spend a sabbatical with his group in the fall of 1989–1990, and he said yes. I hoped to get back to being a scientist again, studying noctilucent clouds with Gary Thomas of Charlie's group. I spent most of my time in Boulder doing just that, but I also got interested in finding new ways to encourage students from grades 5–9 to be interested in science and mathematics.

Charlie had launched a few rockets from Poker Flat, and he and his launch teams were among the brightest and best we helped in my years at Poker Flat. Charlie's group built and flew sounding rocket and satellite-borne experiments and Mars probe experiments. His experimentalists designed and fabricated some of the most sensitive multichannel sensors for ultraviolet spectrometers. His specialty when it came to the aurora was the study of nitric oxide, a presumed catalyst for energy transfer between atoms and molecules resulting in emissions that could not be explained in any other way. Charlie pioneered the work experimentally and theoretically.

Poker Flat Director Neal Brown and family at the gated entrance to Poker Flat in June 1981. Left to right: Neal, his wife Fran Tannian, stepson Michael Sweet, son Nathaniel Brown, stepson Steve Sweet, daughter Melody Brown Burkins holding the family dog Puddles, and son Kristopher Brown

Honorary PhD 1997

My eldest son Kris submitted my name for an honorary PhD from the University of Alaska Fairbanks in 1997. I was informed that I had been selected and that they were going to give another one to the owner of *USA Today*. Both of us would receive the blue and gold UAF banner colors to wear with our graduation robes.

My neighbor Mike Kelly was the president of the UAF Board of Regents. Before putting my banner on me, he told everyone of my love of science and how I took pictures of Denali from the property he cleared next door before building his house there. I took the opportunity to thank him and tell the audience that another reason the day was special was that my stepson Steve Sweet was getting his master's degree from the UAF Institute of Marine Science for his work on the ocean currents between Hawaii and Alaska.

Neal Brown (middle) with his son Nathaniel Brown (left) and the University of Alaska Board of Regents President Michael Kelly (right). Neal received an honorary doctorate degree from the University of Alaska Fairbanks, in 1997

My First Picture of Aurora with a Digital Camera

When I came to Alaska in September of 1963, I took pictures of aurora with black-and-white and color film in my tripod-mounted camera and attempted to make notes about settings because I had to send the film off to Kodak for processing. The slides would come back 2 weeks later, and I found it very difficult to correlate my notes on camera settings with the pictures I had taken. As a result, I often shot a lot of expensive pictures at various settings, by guess and by golly, because there were no internal or external light meters in those cameras that could read the low light of the aurora.

I had seen super-expensive digital cameras capture aurora at Camp Denali Wilderness Lodge, on the northern border of Denali National Park, where I spent many of my summers. I had helped people learn how to set their cameras so they got perfectly exposed pictures every time. The prices of digital cameras were dropping rapidly, and the cameras were getting better.

In early April of 2011 a great display of aurora took place over the city of Fairbanks around 2 a.m. I knew my digital camera, which I had purchased back in 2002, could not capture this display. I did not even try to take a picture and instead spent about 45 min just enjoying the display in relatively warm weather. In less than a week, it would no longer be dark enough to see aurora at night.

Neal Brown's first photo of the aurora with a digital camera, April 2011

The next morning, as soon as Alaska Camera opened up, I was there ready to purchase a new digital camera and lens. I asked for suggestions and bought a $700 Canon T3 camera and a $700 Tokina f/2.8, 18-mm lens.

That night was clear again, and the aurora activity was there soon after sunset and good by 10 p.m. We had a whimsical set of bicycle wheels set up to rotate in the slightest breeze on our deck. I put my camera on a tripod and positioned it and myself so I could get a picture of the aurora and the bicycle wheels. The wheels rotated in the warm, less-than-5-mile-per-hour breeze. I could see that my first picture was perfect, with a slight blur of the rotating wheels and sharp images of the aurora. I took a couple dozen pictures.

By midnight I had selected one perfect picture and uploaded it to SpaceWeather.com. Later that day they chose it to appear on their website as the best picture of that day.

In 2016 my son Kris bought me a Sony A7s, one of the first full-frame mirrorless digital cameras, known especially for low-light performance and video capabilities, which I began to use to take color video of aurora using the same Tokina lens.

Poker Flat at 20

In 1989, when I retired from Poker Flat and 20 years after the first rocket left the ground in 1969, 227 major sounding rockets had been launched. Twenty-nine of those missions required recovery of the payload, and in each of those cases, the recovery was successful. In addition, from the Meteorological Rocket Network site, more than 1200 meteorological rockets had been launched. The array of supporting observational platforms grew from a mere handful of the most essential instruments to a cohesive network that included the Chatanika incoherent scatter radar, the mesospheric-stratospheric-tropospheric radar, the wideband satellite ground station X-band and S-band radars, magnetometers, riometers, meridian-scanning photometers, all-sky cameras, and an elaborate optical spectrophotometer.

Many eminent scientists and research agencies and institutions have used the range. These include, among many others, Jim Gray and John Benevento of NASA Wallops, Norm Peterson of NASA Goddard, Ed McKenna of Air Force Geophysics Laboratory, Gil Moore of Thiokol, Ed Allen and Steve Fisher of Space Data Corporation, Ed Butterfield and Jim Carver of White Sands Missile Range, Bill Barton of Sandia National Laboratories, Herb Mitchell of Research and Development Associates (formerly of DNA), and Chuck Code of Stanford Research International. Several scientific breakthroughs occurred as a result of experiments performed at Poker Flat, which is still today recognized as the world's best-equipped aurora zone research facility.

Aurora researcher reflects on career

Brown leaves long list of successes

By ANDREW ARENSON
For the News-Miner

Little did Neal Brown know that discoveries he made at NASA would lead directly into the heart of the aurora borealis.

SPOTLIGHT

Brown got his first big break in science when he went to work for NASA after graduating from Washington State University in 1961 with a bachelor's degree in physics.

Brown will retire in June after 25 years with the University of Alaska Fairbanks.

Brown's resume tells of a life of hands-on cutting-edge space research.

His accomplishments include being full-time director of UAF's Poker Flat Research Range from 1971 through 1989, one of the nation's busiest space research facilities—the world's only university-owned rocket range.

At NASA's Ames Research Center in Mountain View, Calif., he became part of a group researching ways to get astronauts safely back to Earth.

Brown said getting them into space was not the issue, getting them back through the reentry process alive was the program's biggest problem.

"We could get them up," he said, "but they'd melt when they came back down."

While working for NASA, Brown found what he described as one of the true loves of his life—the aurora, which was a link in understanding the Earth's atmospheric makeup for spacecraft reentry.

As a research assistant and professor at the UAF Geophysical Institute, which oversees operations at Poker Flat, he helped produce the video "The Aurora Explained." He has been on national TV on several occasions including PBS's "Newton's Apple," the "Arctic Light" program on The Discovery Channel, and "Good Morning America." He has also traveled around the country to lecture on the aurora and Poker Flat research.

In February 1993 Brown conceived the idea for the Alaska Space Academy, a week-long summer camp for middle school students.

He believed so much in the camp that in the first two years of its existence, he gave $11,000 of his own money to ensure the camp had the equipment and supplies it needed.

More than 110 students from Alaska, Canada and Russia attended the first summer session in 1993 at Lost Lake, 50 miles south of Fairbanks.

At the camp, students had the opportunity to scan the floor of Lost Lake with a remote-controlled submersible camera, mapping the lake as if it were another planet. They built and launched rockets, then entered their data into computers.

Brown planned camp activities in such a way that science and math were shrouded in fun activities. Not only did the students have a great time, they learned science is fun—Brown's ultimate goal.

"I like to work with kids," Brown said. "I always got high marks for communicating with them. The space camp was the most rewarding thing I've ever done—bar none. I argued we could start a summer space camp that kids would want to come to—and I was right!"

Brown has been doing aurora research at Poker Flat for the better part of 20 years. He claims that over the years, he has logged nearly 500,000 miles driving the Steese Highway between the range and the Geophysical Institute, roughly the distance to the moon and back.

Although he never left the ground, he has seen some amazing things at Poker Flat, not all of which were happy circumstances.

"The worst thing I worked through was in March of 1974 while we were loading the biggest rocket to date," Brown said. "It was huge, bigger than anything we'd done."

The crane that held the rocket chambers broke, dropping the rocket and its payload on the launching pad, crushing the foundation and rocket.

"Inside the wreckage were seven 30-pound canisters with very unstable explosives," Brown said. "The person who knew most about this type of situation was severely injured in the accident because the cone of the rocket had struck him so hard on his head that it fractured his skull and shattered one of his leg bones.

"We took charge of the situation and flew a specialized team up from New Mexico within 24 hours to help advise us."

Neil Davis, one of the founders and former director of Poker Flat, recalled the situation as critical because of the explosives and barium used in the rocket.

"The payload was fractured and the barium could have easily set off an explosion if it got wet," Davis said. "We met with the team from New Mexico and came up with recommendations. We sent one man out—an explosives specialist—so if anything did happen, only one person would be lost."

The situation was defused and the explosives stabilized. It took three days to clean up the mess.

Davis and Brown developed a strong friendship.

Davis said that while he was director of Poker Flat, he would dream up awful scenarios during which he could pick only four people to help figure out the fictitious problem.

"Neal Brown was always one of the people I would pick," Davis said. "Neal was extremely good in clutch situations; he was always very cool, confident and collected."

The range hosts researchers from all over the world to study the aurora. Brown said Poker Flat is unique because it is run through UAF and is exclusively a research station.

However, Brown laments that unlike earthquake studies, space science is not as "entertaining" to people as it once was.

"How do you tell the public about this exciting stuff?" he asked. "Everybody can understand science, but space has not been able to come up with the same exciting stuff that seismology has."

He looks toward his retirement as a great opportunity to work on other projects, such as building science exhibits for museums and speaking to kids about the aurora. He also plans to start writing the history of Poker Flat and will continue helping with the space camp.

Andrew Arenson is a senior at the University of Alaska Fairbanks majoring in journalism.

Jim Seida photo

ROCKET MAN—Neal Brown will retire from the University of Alaska Fairbanks this June after 25 years of space research often focusing on the aurora borealis.

An article in the *Fairbanks Daily News-Miner* upon Neal's retirement, July 10, 1989

Through the years, Poker Flat has been sponsored by a variety of federal agencies, but never by State of Alaska dollars. Among the sponsors are the National Science Foundation, Sandia National Laboratories, the Geophysical Institute, Advanced Research Projects Agency, and above all NASA, which funded about 43% of all efforts up to my retirement, and the Defense Nuclear Agency, which funded about 38%.

The annual operating budget and the permanent staff have greatly fluctuated from year to year, from a mere $500 with no employees after the first launches to $1.2 million and 50 people at the peak of construction to approximately $750,000 and six staff members in 1989 when I retired. The range is able to support from 15 to 20 major sounding rocket launches a year, which is quite a full schedule.

Poker Flat has a long history of cooperation and coordination, not only with its users and sponsors but also with the various support groups. Each launch is carefully coordinated with the FAA, which grants permission to use air space and assures that it is clear of traffic during launches. The site itself is on State of Alaska land on a long-term lease to the University of Alaska. Permission has been given by the federal government to impact its lands to the north of the launch site.

The importance of the role played by the Atmospheric Sciences Laboratory meteorological team from Fort Greely, Alaska, which runs the MRN site, cannot be overemphasized, nor can that of the Alaska Air Command. Through my retirement in 1989, the 709th AC & W Squadron at Fort Yukon not only tracked recovery payloads but also supported scientists using the Fort Yukon field site. The 5010th Helicopter Squadron made the actual recovery possible. In addition to these services, both the army and air force provided Poker Flat with assistance in innumerable other areas.

Poker Flat's status did not happen overnight, nor was it achieved by the efforts of any one group of individuals. There were times, especially following that first launch season, when it was doubtful if there would even be a Poker Flat Research Range.

And I end with this quote:

> *No one goes his way alone; all that we send into the lives of others, comes back into our own.*
>
> *—Edwin Markham*

Poker Flat Songs and Poems

Often in the long midnight hours when we were waiting for the right launch conditions, a voice would come over the intercom system to announce that they had a song or poem to offer. Here are a few of those late-night ramblings:

Quasi-Official Icecap '75 Rocket Range Song

By the Chatanika Radar Chorus and Orchestra (all rights reserved)

Sung to the tune of *Sloop John B* (our apologies to the original author)

Verse 1:
Oh—we come to the Poker Range
The people are very strange
We come to shoot the rockets into the sky
Jim Ulwicks's the boss
He's afraid of a loss
I see the break-up now—I wanna go home.
Chorus:
So hoist up the (Loki) Dart (Niki, Sergeant, Saturn)
Set in the az and el
Push the button quick to make it go away
Let me go home
Please don't let the damn count continue 'til May.
Verse 2:
The Pad Chief he got drunk
He went to find his bunk
Instead he fell asleep on Launch Pad Four
The aurora it came
And no one's to blame
Poor Carroll Coe got launched with a hellova roar.
Verse 3:
With T minus 3 and hold I wish for one man bold
To push the button and make the rocket go by
Come on Baker and Stair
Don't be so damn square
Count it down and shoot—I wanna go home.
Verse 4:
The rocketeers are sad
The road block boys are mad
The payload's chilled and all the systems are go
But where's Neal Brown
No where to be found
He's off the road and his car is stuck in the snow.
The Baffled Field Widen Anthem
Sung to the tune of *McNamara's Band*
Verse 1:
Oh, our name is field widen, we're the cream of Utah State
Just ask about our instruments, we'll tell you that they're great. Our latest one is baffled, (and often so are we)
We hope our coming flight succeeds, (so far, they're data free)
Chorus:
Oh, the springs, they twang, and the slide goes bang, and we hope it doesn't jam,

Cause DNA will only pay for interferograms.

We're nursing our computer, and we pray it doesn't crash cause scientific data's nice, but Utah needs the cash

Verse 2:

We come to integrations, and to rocket launches too!

We charge the biggest bucks because we field the biggest crew.

We're Baffled Field Widen; our cryo tanks are full.

We'll dazzle you with data, or baffle you with bull.

Verse 3:

If you would like to save us, Jim, from still another poem

For God's sake, shoot this rocket, man, and send those people home!

Cause if you don't, I promise that tomorrow will be worse. I'll even save this piece of trash, and add another verse.

Last Night's Threat Made Good

Let's hear it for Jim Ulwick, our Massachusetts Ute, Who, gallantly, refused a twenty-dollar bribe to shoot

His face showed pain, and great disdain as he refused the twenty, and he said "Scott that's quite a lot. I think fifteen is plenty."

The Ballad of Spirit and Hiram

(or we've got good news, and bad news)

This poem is not a happy one, but if you choose to hear it

Read on, my friend, and learn about the troubled times of SPIRIT.

She gleams like burnished silver on the payload building floor

But she may run out of window, before rolling out the door

Her five detectors aren't wine, old age does not improve them

Sometimes they work, sometimes they don't. Just when "the spirit moves them."

Her gyro died of cause unknown; a relay passed away "progress avoidance in effect" was the motto of the day.

She been undressed more times than any Lonely Lady stripper

The next revision won't have screws, instead they'll use a zipper.

But things are looking up they say, and we pray for a decision to put the skin back on for keeps - with no more circumcision.

The hallmark crew will pull her through, and Thursday night they'll try

to hang her on the rail again, and point her at the sky

Let's bow our heads, and clasp our hands and say a prayer for SPIRIT

that her demons have been exorcised, and never more go near it And while we're at it, say one for continued health of HIRAM

it's nice to get 'em on the pad, but nicer still to fire 'em!

The Birth of Poker Flat

(From *Tall Tales of the Tundra*)

A little piece of tundra fell from out the sky one day

And landed in Alaska, right beside the Steese Highway

The Rocket Gods who found it thought it looked so cold and strange,
They knew it wasn't good for much, except a rocket range.
They stuffed it full of launchers and cocoons of Styrofoam (the only place you'll find them, no matter where you roam.)
And when the pads were done they built a blockhouse at the hub (so users have a place to sit each night until they scrub)
A hilltop road was built to use for sledding recreation (which, incidentally, goes to Optics, and the TM station)
For winter sports enthusiasts, they made a place to stay the Poker Inn! (It's modeled on the Hilton in L.A.!)
They built a payload building for checkout and assembly
(Space Data built the benches, if you wonder why they're so trembly.)
It gives the users space to work, with shelter from the weather
And it's better than a circus, with three groups there together.
They knew they needed radar, and the one they have's a winner
It's backup for the microwave, to heat your TV dinner.
With chassis, wheels, and engine parts, some metal, pine and spruce
They made the damndest car there is, and christened it "The Moose."
The rocket launching business hasn't paid its way for years
They make their real money from the sale of souvenirs
They have a so called "office" but it's not the way it looks.
It's just a front, the big bucks come from T shirts, hats and books.
Now Rocket Gods are much too proud to run the range themselves
They need a crew to do the work (like Santa needs his elves). They also need a manager (someone to take the blame).
And last, but hardly least, they had to give the place a name.
So they shanghaied Carroll, Steve, and Ed, and Mike, Rick, VInce and Pat
They named Neal Chief of the Arctic Fief, and they called it Poker Flat.
—Ron Newcomb 4/83

Just Another Night

Here I sit another cold night
Waiting for that northern light
my camera ready, and gloves in hand waiting for that spirit clan
Then there's Ernie, who said to me
Be patient young one, you're still so green
There will come a time, for which you'll see
Those northern lights from tree to tree
Aurora Borealis I said to thee
Come and shine your lights on me
For when you do, we'll all be free
To get back in the saddle with our Q-tee
Like Prancer and Dancer up in the sky
It was McKenna and O'Connor flying high

But there were no lights on this blurry night
And Patricia said with all her might
We'll see you tomorrow to do it again
Drive 'o so careful
and stay out of shenanigans
—Mitch Thompson, 1/86

Just FWIF!

As people dashed for the blockhouse door
The rocket was launched with a deafening roar
The Field Widened Payload was launched at last
And worries and fears were now part of the past
The slide really worked, the way it was planned
and the cry of the crew was "interferograms"!
The aurora was waiting as the rocket roared through as if it knew what we wanted it to do
As data poured in and the excitement grew
XXXXXXXX up we're almost through
All faces were smiles as we broke out champagne
Cause we'd just won the Kilorayleigh Game.

Oklahoma

(Rodgers and Hammerstein want to make it very clear that this has no connection with their show.)

We all have crosses we must bear, the Christians had the Romans
The Turks had their Armenians, and we have Oklahomans
Don't get us wrong, we're grateful for their stocks of gas and oil
We got stuck, confound our luck, and we have Dean and Doyle
We all know that they're Preeverts, cause we hear them at their station
And Doyle is always asking Dean about his "deviation"
The kinky things they talk about are proof of their pree version
Like "turning on" and "turning off" and "A to D conversion"
But morals and behavior are none of our affair
cause they really shine at countdown time when TM's on the air
So let us all be tolerant, no matter if they're strange
and make the Okies welcome, and at home upon the range

[Our Okies were Dean Feken and Doyle Craft and they used the offbeat words to describe operations of their radio telemetry, or TM, equipment.]

The Plight of the Northern Lights

(or *A Poem for the Rocketeer*)

Listen my friends and you shall hear of the midnight ride of a solar flare
On the sun far away this flare is born beginning its life in a solar storm
Passing through space with incredible speed, creating a phenomonum the scientists need

This condition known as the Northern Lights, really creates a phenomonial sight

Rockets are readied and people wait, to catch this energy in its natural state

Tension is high when there in the dark

is the evidence of an auroral arc

The countdown starts with all systems go, often ending with a conditional hold

Recycle the count and hope and pray,

the aurora will come back without delay

Finally amid the gloom and despair,

the perfect arc is finally there

The rocket is launched with a deafening roar as on to the auroral arc it soars.

Data is taken at a fantastic rate

to get the best info on aurora to date

Everything works and up goes the cheer

of the crew that supported this rocket this year

To think this all started with a solar flare

and a handful of people beyond compare

I think even when the aurora is then, we'd all be willing to do it again!

—Anonymous

The Project Scientist, He Really Does Exist!

On a hilltop high, 'neath a cold, black sky, shrouded in legend and mist,

Sits the man of the hour, in his ivory tower: Our Project Scientist.

Each night we wait, for he holds our fate in the palm of his powerful hand

Each wave or nod, like a word from God. His wish is our command.

With slackened jaw, we watch in awe, as he, so wondrous wise.

Searches the dark for a small faint arc in the starry polar skies

Down here below, we mortals know that in his hallowed dome

Our hero is infallible, just like the Pope, in Rome

He carefully weighs the peaks and bays from magnetometers, and well defines the spectral lines from range photometers

He wades through stacks of published facts in reaching his decision

with iron nerves, he never swerves from clear objective vision

Some common folks make cruel jokes about his rubber windows which grow or shrink, these cynics think, as fickle as the wind blows

But stones, nor sticks, nor politics, nor threats, nor strong invective

Can sway our man from his master plan, nor blur his sharp perspective

Despite the odds and angry Gods, He'll fly his sacred rocket

The magic means are in in his jeans, tucked safely in his pocket

So when it's time to get off the dime, our guys won't pass the buck

He'll stand up tall and make his call, and cross himself for luck

He'll never wait or hesitate when it's decision time

with crafty gin, he'll reach far in, and grab his lucky dime
With practiced grip, and skillful flip, he'll toss the coin up high
If it lands, a tail, He'll hug the rail; If it comes down heads, He'll fly
So all you guys with bleary eyes, who's like to see your beds wish not for-a, great aurora, pray for heads, instead
And here's to you, oh Great Guru, thou Prince of Scientists,
Each night's delay assures you'll stay on everybody's list
Which list I fear, I can't say here; it cannot be repeated
But, be advised: This poem's revised, with expletive deleted.

Newcomb 1986 St. Patrick

This the night of St. Patrick, at old Chatanika
The intercom's quiet, no sound from the speaker
The launchers are pointed, with care, at the sky
In hopes that it's this night the rockets will fly
The conjurer's eyes burn a hole in the screen
But nary a trace of aurora is seen
While Lynch strokes his beard, and Scott fondles his hat
The optics site tells us the mags are all flat
TM is all ready; the payloads are go
But clouds have rolled in, and its starting to snow
Fort Nelson's socked in, Peace River is closing
The Hallmark crew's all watching movies or dozing
The chances of seeing Aurora Borealis
Are lower than snakeskin (just like our morale is)
ELIAS gives up; Field Widen goes prime
Jim Ulwick hangs in till he runs out of time
It's been a tough nite; never once went internal
But a rocketeer's weird, his hope springs eternal
And although tonight, we saw no aurora
Nobody gives up, cause there always tomorra
—*Anonymous*

Epilogue: Neal Brown's Legacy

Neal Brown retired as director of Poker Flat Research Range on June 30, 1989, having overseen the range through enormous challenges and growth. But Neal did not fully retire until years later. He continued to do education outreach with both students and teachers about the aurora and space and give lectures on these topics as well as physics and unmanned aircraft. Until he finally left Alaska in 2019, he was the go-to person when anyone requested a scientist to talk about the aurora. From 2002 to 2008, he was director of the Alaska Space Grant, a program funded by NASA in every state with the intent of getting undergraduate students hands-on experiences and internships at various NASA and aerospace industries. Neal established and ran the Alaska Space Academy for middle-school students for two summers, helped develop curriculum for teaching about aurora, taught model rocket building, and worked with Iñupiaq Elders and knowledge bearers and Geophysical Institute scientists to create the video *Kiuguyat: The Northern Lights* in both a flat-screen version and planetarium-dome version. The film was shared with teachers, students, and communities in the North Slope, Nome, and Northwest Arctic Borough School Districts.

Neal was on the board of the University of Alaska Museum of the North for many years and worked to establish a planetarium there, a goal that finally came to fruition in 2026. Neal's advocacy helped the Geophysical Institute to acquire a portable planetarium that has been taken to schools throughout the state.

Neal also spent time in the summers at Camp Denali, a wilderness lodge deep inside Denali National Park, giving an after-dinner program and talking informally with the guests about aurora and visible atmospheric phenomena such as sundogs and haloes. When it was dark enough to see and the aurora was out, Neal would wake the guests who wanted a chance to see the aurora on their visit.

Through all this, Neal practiced taking photos of the aurora, often staying out late at night. He used to say that the best way to ensure you saw the aurora was to drink a large glass of water just before bed. As Neal said in the video *Aurora Explained*, "To this day, if there is a spectacular aurora, I'll stop dead in my tracks.

N. Brown, *Northern Lights and Rocket Flights*, Springer Biographies,
https://doi.org/10.1007/978-3-032-14598-7

I will pull over to the side of the road. In fact, a couple of years ago, I pulled over to the side of the Steese Highway at 2 am and just shut the car down and laid down in the road and watched the aurora, because it was so incredibly beautiful."

Neal's enthusiasm and ability to clearly explain difficult concepts made him an excellent teacher, and he loved when he could spark someone's interest so they would learn more on their own.

Neal wrote:

I think of myself as a teacher and a father and take some pride in doing those things well. I have developed some listening skills for what my students, my family, and my friends are telling me. I listen and try to answer them at their own level. Sometimes on purpose I tweak my response to get them interested to learn more on their own. I love to teach.

I teach without expecting praise from my students. I'm glad I helped them understand that they know more science than they think they do, and that ultimately they, and I, are both lifelong learners, which seems like worth investing their time in.

I have come to realize how important stories are that inspire the imagination. I have used stories to get my teaching points across without thinking about this imagination thing, I think. Instead I have routinely told about experiences from my own life that I think will connect with experiences in my students' lives to teach my lesson or to get my point across.

By crafting your stories about the things you are interested in, and telling them in concise ways to your students with a hook, explanation, and enthusiasm, you will connect in those wondrous ways teachers and mentors do. Your students will enjoy your story, they will create their own imagery to go with your story, and perhaps without even knowing they have learned, you will have set them on a path to using their imagination—perhaps even to using their imagination to create a future they want to take part in.

Neal said that when former colleagues reminisced with him, hardly anyone asked about the one thing he spent the most energy on and that meant the most to him—and that was being a father to Kris, Mel, and Nat, and then stepfather to Steve and Mike after he married his beloved wife and, as he often said, best friend Fran.

Neal wrote of himself, "If I call someone who does not remember me, I sometimes just say, 'I'm the big guy that wore suspenders when I gave a talk to the group you were in.' My wife Fran uses the same thing in a department or grocery store to ask if anyone has seen her husband. 'He's a big guy and always has suspenders on.'" Neal ended with this quote by Edwin Markham: "No one goes his way alone; all that we send into the lives of others comes back into our own."

In 2019, after 55 years in Alaska, Neal and Fran moved to New Hampshire to be closer to family. Neal passed away there in 2021. Following his death, his family worked with the UA Museum of the North in Fairbanks and the Geophysical Institute to set up the Neal Brown Museum Education Endowment to foster scientific learning and curiosity.

100%

POKER FLAT

BY
KRIS BROWN

NOVEMBER 29, 1981

①

I am interested in Poker Flat. Poker Flat is a privately owned research range located thirty miles north of Fairbanks on the Steese Highway. Poker Flat is a launching site for rockets which are sent up into the upper atmosphere to conduct experiments. My father, Neal Brown, is the range supervisor and I have asked him if I can interview him this coming Saturday. He agreed, but suggested that I should also try to interview one of the people who work at Poker Flat. I think that that is a good idea but it's going to be a lot harder to interview a stranger than my dad.

What should I do? I have checked out two books on rockets and payloads and have taken some notes which I will use to explain where and how the experiments which are conducted at Poker Flat are done. I am trying to learn about how everything at Poker Flat works, so that I will be prepared for my two interviews. I have found that Poker Flat is primarily used for unclassified type research, but does conduct a few types of experiments which are almost classified. This information is quite interesting, and I would like to learn more on the subject. I will be sure to include a question on the topic when I do my

(2)

interviews.

The day of my first interview, I wasn't quite prepared. I had to think of a lot of questions to ask my father and, I really wasn't sure what to ask. After finally tracking him down in his bedroom after lunch, I started my interview. My father gave me a general idea of what Poker Flat is used for and then briefly described to me how some of the experiments are conducted. I listened attentively to the information which I had pretty much already known through my past knowledge of Poker Flat and then got to my questions.

"Who owns and operates Poker Flat?"

"The University of Alaska's Geophysical Institute both owns and operates Poker Flat."

He also added that Poker Flat is the only research site of its type in the world and that it is the only research site in the United States where experiments on the aurora borealis can be done.

I was curious as to where they received the money which is needed to run such an installation. "Who does Poker Flat receive its annual income from?"

"It is funded by three major government agencies. The first being NASA,

③

which provides one-half of the total cost for running Poker Flat. The second is the U.S. Air Force, and the third, the Defence Nuclear Agency."

He continued on to tell me exactly what Poker Flat is used for. He told me that the experiments at Poker Flat were done to learn more about the aurora, its effect on the world, and on military type subjects such as infared radar. The range, he said, is located just south of the auroral oval, and is thus able to receive a large part of the total auroral activity which occurs each year. It is through this strategic location that such experiments as the study of bariums effect on the atmosphere to produce artificial aurora can be done.

He answered a few more of my questions and I copied what information I could down. I thanked him and then went down to my room to write some of it in a fashion which I could more easily understand.

I looked over my notes the next day and was surprised to find that I really hadn't discussed the actual process of conducting an experiment. I would be sure to ask Pat Young, the range launch control

④

officer, when I interviewed her this coming Tuesday.

I also was surprised when I added together the entire amount of people which were needed to launch a rocket and conduct an experiment. I found that a launch requires: one scientist, two rocket technicians, three people for telemetry, three for radar, one to launch the rocket, one for the countdown, one electronic technician, and one person to supervise the whole affair. There are also several people who are needed to maintain the range itself. All together, there are approximately twenty people on the range during an experiment. I found this quite a few, considering that the range is quite small and doesn't have a lot of room.

I have arranged to talk with Pat Young on the telephone this Tuesday after school. I am fairly well prepared for the interview and I am sure that I will have no trouble with it

I came home after school today and after a quick dash to the refrigerator dialed the number of Poker Flat. Pat Young answered the phone. I told her about why I had called and inquired as to if I could ask her a few questions. She said it would be okay, so I began.

(5)

I was wondering about the launch procedure and what was needed in order to start an experiment. She answered the latter of my two questions first by running through the list of events which have to be done for an experiment to occur. The procedure is as follows: A proposal is sent in by a group of scientists to a government agency such as NASA. NASA reviews the proposal and if it seems worthwhile gives the scientists money. The scientist then ask Poker Flat to find a rocket (military surplus). The scientists then build a payload for the rocket which holds their experiment. A contract with Poker Flat is now signed to insure Poker Flat receives enough funds. After the contract is signed, the scientist come to Alaska, assemble the rocket and payload, launch the rocket, and spend about a year examining their data. This whole procedure takes approximately four years.

I was happy to finally be able to understand what all was needed for an experiment to occur, but what I really wanted to know now was how an actual launch was done. This was my next questions.

"A twenty-four hour ~~is~~ notice is given to the F.A.A. and the Alaska Air Command, before each launch. The F.A.A. and the

⑥

Fairbanks International Airport are warned again three hours before the launch. An hourly notification is then given after that until T-thirty minutes. The F.A.A. and the airport are then called every five minutes. At T-ten minutes, the airport is told to stop all air-traffic over Poker Flat for approximately twenty to thirty minutes. At T-one minute the time between each launch notification is ten seconds until T-ten seconds, when each second starts to be counted. If everything goes well, at T-zero seconds, there will be a lift-off."

I could not believe how much was needed to send a rocket up. It was incredible. I thanked Pat Young for her help and she asked me if I would be interested in talking to Mike Cogen, the range wind waiter. I replied with a yes and then waited patiently on the telephone as she paged him.

Mike Cogen's job is a wind waiter. This means that he is in charge of keeping track of how the weather conditions are and how they might affect the launch. His main task is dealing with the wind, although he does have to make sure that the proper conditions are present for each experiment. For example, if an experiment is done to learn about

7

sunspots, and the sun doesn't seem to have any sunspots on it the day of the launch, there might be a little trouble. Mike Cogen's job is to make sure to have, such a launch on a night when there are sunspots.

Mr. Cogen studies the wind and its affect on rockets. By using balloons, Mr. Cogen can find the speed of the winds, and by feeding this information into a computer designed for such work can determine the correct angle needed in which to point the rocket, to allow it to reach its target.

I asked Mr. Cogen after our conversation if there was anything he had to do once he had given the information to the computer. He replied with a chuckle that all he did was sit back and watch.

My search about Poker Flat is basically over now, but I still think there is a great deal more that I would like to know. I feel I have learned a lot and will always remember this information.

References

I. G. Edmonds, Jet and Rocket Engines How they work, New York, G.P. Putnam's Sons, 1973, 71, 72, 73

Dr. Wernher Von Braun, Space Frontier, New York, Holt, Rhinehart and Winston, 1963, 27-28

Interviews

Neal B. Brown, Range Supervisor for Poker Flat research range, Geophysical Institute, University of Alaska November 21, 1981

Pat Young, Launch Control Officer at Poker Flat research range, Geophysical Institute, University of Alaska November 23, 198[illegible]

Mike Cogen, Wind Waiter at Poker Flat research range, Geophysical Institute, University of Alaska November 23, 1981

Outstanding information
Outstanding format

Neal Brown Obituary

Neal Boyd Brown, 82, died Monday, July 19, 2021, at Dartmouth Hitchcock Medical Center following a brief illness.

Neal was born December 13, 1938, in Moscow, Idaho, the son of Kenneth Wayne Brown and Ruth Alvina (Boyd) Brown. Growing up on the family farm, he learned how to build and fix almost anything mechanical and electrical with whatever was handy, a skill he used—and taught—throughout his life and which earned him his Eagle Scout ranking in the Boy Scouts. From an early age, Neal said his parents nurtured his curiosity to become a lifelong learner. It was Neal's curiosity, his love of life, and the joy, laughter, and kindness he so eagerly, enthusiastically, and generously shared with all who knew him that is both his legacy and our blessing.

Neal graduated from Pullman High School class of 1957 in Pullman, Washington. He went on to earn a bachelor's degree in physics from Washington State University, a master's in geophysics from the University of Alaska Fairbanks (UAF), and later a legislative citation from the Alaska State Legislature (1995) and honorary doctorate from UAF (1997) for his contributions and service to Alaska.

Neal spent most of his professional life studying, photographing, and teaching about the aurora borealis, or northern lights. Seeing them first in 1962 during a NASA research posting in Thule, Greenland, the "lights in the sky" were a source of delight and wonder for him throughout his 18 years (1971–1989) supervising and directing UAF's Poker Flat Research Range, the world's first and only scientific rocket launching facility owned by a university. The aurora continued to inspire him as he was asked to take on the position of director for Alaska Space Grant from 2002 to 2008. In that role, he most enjoyed working with Alaskan teachers and remote communities to offer STEM education summer field camps to young Alaskans, emphasizing outreach to under-represented, rural, and Indigenous learners. In between building and launching model rockets, his young students explored Alaska lakes with remote fish finders, simulating the exploration for life on distant planets by NASA remote sensors. After retirement, he continued to teach about Alaska

N. Brown, *Northern Lights and Rocket Flights*, Springer Biographies, https://doi.org/10.1007/978-3-032-14598-7

science and the aurora to both academic and public audiences whenever asked, as the founder and chief scientist of Alaska Science Explained.

Neal's ease in explaining complex scientific concepts to public audiences; his charismatic, gentle, and kind character; and his wonderful storytelling made him a frequent guest for interviews on National Public Radio and other national and international media outlets. He has also appeared in several videos that showcase one of his proudest co-creations with his colleagues Tom Hallinan and Dan Osborne, the Aurora Color Television Project—the first broadcast-quality, color television footage of the aurora, created in 2008. Not only did he love the complexity and beauty of outdoor photography; he also loved to teach others how to both photograph and film the northern lights and other natural phenomena. He spent time as a mentor to others while he continued learning as much as he could to keep up with the rapidly evolving technology of digital cameras and camera-carrying drones. These ideas were woven into his teaching about the aurora, science, rockets, and space whenever asked, be it at a local school or Osher Lifelong Learning programs and community groups in Alaska, New Hampshire, Vermont, and more. He also spent countless Alaska summers giving lectures on the aurora borealis to audiences from around the world at Camp Denali, a wilderness lodge in Alaska's Denali National Park. He often joked he spent his life "learning about things over his head."

In addition to his curiosity about the Alaska night sky, the aurora, space science, and photography, Neal was an avid genealogist and ham radio operator. Both activities let him share connections, ideas, and his joy of learning and discovery with people around the world. He spent over 20 years researching the lives of not only his family but the entire families of his children's spouses, finding historical relationships and family stories many had never known. Through ham radio, he enjoyed being part of a world connecting people through electronics and stories.

No story about Neal Brown would be complete without noting his love of the movie *Star Wars*—a movie of space exploration, good triumphing over evil, and connections across the universe. He and his children watched the original together over 20 times when it first came out in 1977, and Neal not only made his own Darth Vader costume that year for Halloween, he also built a moving, beeping, rotating, remote-controlled R2-D2. He wore his costume and brought R2-D2 to visit classrooms, local parades, and community parties whenever asked for many years to come. Star Wars movie clips, quotes, and theater visits with friends and family to watch the sequels brought him immense joy.

His generosity in sharing knowledge, creating fun, and experiencing wonder knew no bounds. As all who ever met him knew, a simple conversation on the street might lead to years of correspondence and the sharing of ideas, jokes, family events, and more. He loved connecting people to one another and seeing the good in everyone, maintaining his many friendships through regular letters, emails, social media postings, and phone calls as well as old-fashioned visits for lunch or dinner whenever he was nearby.

With all his work and hobbies, however, Neal's foremost love was for his wife of 40 years, Fran Tannian, and their family. He loved to surprise Fran with bouquets of fresh flowers, often daisies, and adored the quiet time they spent with their little

Bichon Frise dog, Molly, watching birds outside their window and catching up on their day. He also loved to dance with Fran, travel with her, share jokes with her, and appreciate how well they took care of one another as they both got older. For his children, grandchildren, and extended family, he was a daily presence in letters, emails, photo sharing, and social media. He loved connecting them to one another, keeping them updated about each other's lives, and making sure birthdays, anniversaries, and life events were remembered and celebrated. Few days went by that he would not send cartoons, newspaper stories, pictures, and books to each of them, sometimes in duplicate, whenever he found something he thought they would like. He kept the US Post Office busy each week with multiple mailings to family.

After enjoying over 55 years living in Alaska, Neal and Fran moved closer to his daughter and her family in East Thetford, Vermont, in the fall of 2019. Neal and Fran immediately made a community of new friends and, though they missed Alaska, were actively enjoying their lives at the Woodlands retirement community in Lebanon, New Hampshire. Neal not only kept up his public lectures, storytelling, photography, and family updates; he took great joy in learning how to draw and had filled several notebooks with sketches and paintings. He had also started teaching model rocket classes to other residents of the Woodlands.

Neal leaves behind his wife, Fran Tannian, and the children of his first marriage with their spouses, Kristopher David Brown and Rachel Moritt Brown of New York City, New York; Melody Brown Burkins and Derek Logan Burkins of East Thetford, Vermont; and Nathaniel Scott Brown and Tina Tsiakalis-Brown of Seattle, Washington. He also leaves his two stepsons and their spouses, Steven Ross Sweet and Iva Neveux of Reno, Nevada, and Michael Scott Sweet and Laura Pelton Sweet of Boston, Massachusetts. And he leaves six beloved grandchildren: Ari Moritt Brown and Noa Brown of New York City, Riley Logan Burkins and Porter Brown Burkins of East Thetford, and Zacharias Tsiakalis-Brown and Michael Tsiakalis-Brown of Seattle.

Memorial contributions may be made to one of Neal's favorite places—the University of Alaska's Museum of the North—to support the continued sharing of Alaskan science, knowledge, storytelling, and wonder.

Index

N. Brown, *Northern Lights and Rocket Flights*, Springer Biographies,
https://doi.org/10.1007/978-3-032-14598-7

www.ingramcontent.com/pod-product-compliance
Lightning Source LLC
Chambersburg PA
CBHW061415100826
49614CB00003B/28
9783032145970